How to Pass

SECOND EDITION

HIGHER

Physics

Paul Chambers
and Douglas Gavin

HODDER
GIBSON
AN HACHETTE UK COMPANY

The Publishers would like to thank the following for permission to reproduce copyright material:

Photo credits

p.10 © James Steidl – Fotolia.com; **p.12** (left) © Joggie Botma – Fotolia.com, (right) © Erni – Fotolia.com; **p.21** © TRL LTD./SCIENCE PHOTO LIBRARY; **p.77** © ANDREW LAMBERT PHOTOGRAPHY/SCIENCE PHOTO LIBRARY; **p.79** © ANDREW LAMBERT PHOTOGRAPHY/SCIENCE PHOTO LIBRARY; **p.80** © EDWARD KINSMAN/SCIENCE PHOTO LIBRARY; **p.91** © Phil Degginger / Alamy; **p.133** © atm2003 – Fotolia.com; **p.134** © dan talson – Fotolia.com.

Acknowledgements

Questions, where marked throughout by asterisk, are used by permission Copyright © Scottish Qualifications Authority. All questions, answers and worked examples without an asterisk do not emanate from SQA material.

Every effort has been made to trace all copyright holders, but if any have been inadvertently overlooked the Publishers will be pleased to make the necessary arrangements at the first opportunity.

Although every effort has been made to ensure that website addresses are correct at time of going to press, Hodder Gibson cannot be held responsible for the content of any website mentioned in this book. It is sometimes possible to find a relocated web page by typing in the address of the home page for a website in the URL window of your browser.

Hachette UK's policy is to use papers that are natural, renewable and recyclable products and made from wood grown in well-managed forests and other controlled sources. The logging and manufacturing processes are expected to conform to the environmental regulations of the country of origin.

Orders: please contact Hachette UK Distribution, Hely Hutchinson Centre, Milton Road, Didcot, Oxfordshire, OX11 7HH. Telephone: +44 (0)1235 827827. Email education@hachette.co.uk. Lines are open from 9 a.m. to 5 p.m., Monday to Friday. You can also order through our website: www.hoddereducation.co.uk. If you have queries or questions that aren't about an order, you can contact us at hoddergibson@hodder.co.uk

© Paul Chambers, Douglas Gavin 2019

First published in 2019 by
Hodder Gibson, an imprint of Hodder Education,
An Hachette UK Company,
50 Frederick Street
Edinburgh, EH2 1EX

Impression number   6

Year   2023

Cover photo © gazanfer – stock.adobe.com
Illustrations by Aptara, Inc.
Typeset in 13/15 Conos Pro (Light) by Aptara, Inc.
Printed in the UK by CPI Group Ltd
A catalogue record for this title is available from the British Library
ISBN: 978 1 5104 5236 7

SCOTLAND EXCEL

We are an approved supplier on the Scotland Excel framework.

Find us on your school's procurement system as

Hachette UK Distribution Ltd or *Hodder & Stoughton Limited t/a Hodder Education.*

MIX
Paper | Supporting responsible forestry
FSC™ C104740

Contents

How to pass Higher Physics

This *How to Pass* book has been written to help you achieve the best possible grade in your Higher Physics exam. It is written around the information given in the SQA National Course Specification and Course Assessment Schedule. The National Course Specification states that as a result of following a Higher Physics course, candidates should:

* develop and apply knowledge and understanding of physics
* develop an understanding of the role of physics in scientific issues and relevant applications of physics
* develop scientific inquiry and investigative skills
* develop scientific analytical thinking skills, including scientific evaluation, in a physics context
* develop the skills to use technology, equipment and materials safely, in practical scientific activities
* develop planning skills
* develop problem-solving skills in a physics context
* use and understand scientific literacy to communicate ideas and issues and to make scientifically informed choices
* develop the knowledge and skills for more advanced learning in physics
* develop skills of independent working.

This book will assist you in acquiring the skills, knowledge and understanding required for the course. It has been divided into 15 chapters, each of which will open with the 'what you should know' statements for that topic. It is intended that the book will help you optimise what you have already been taught in class and will assist in your revision and preparation for your final examinations.

In a problem-solving context, the book will present the required relationships in order to answer the range of questions in the examination and will show in detail the steps required for you to achieve the correct answer. It will *not* derive or show where the formulae or equations came from, rather it will justify what relationship should be used and show how to use it in the most productive way so that you maximise your mark.

There will be two open-ended questions in the exam. There are various types of open-ended questions. You could be asked to comment on an analogy, a situation or a quotation. When you answer the question, make that sure you refer to the question. It is not enough to recognise what the question is about and then write down everything you know about the subject. If you don't comment on the analogy/situation, etc. you are in danger of not having answered the question and therefore of being awarded no marks.

Open-ended questions are worth 3 marks. They are marked holistically. A good answer is awarded 3 marks (it does not have to be perfect or complete), a reasonable answer is awarded 2 marks and a limited answer is awarded 1 mark.

Hints and tips boxes are scattered throughout the text to offer specific guidance on a particular topic. Each chapter concludes with a list of the key points covered in the chapter, the important vocabulary for the topic and a number of practice questions to test your understanding. When a question number is preceded by an asterisk (*), this indicates that the question is from an SQA past paper.

Marking principles
General points

- Positive marking: marks will be awarded for correct physics; marks will *not* be lost for errors or omissions (i.e. there is no deduction from marks you have already gained. This does not mean that you will gain marks for wrong or missing physics, however). The unit will form part of the answer.
- Unless working is specifically asked for, a correct final answer (including unit) will receive full marks. If a question asks you to 'show' how you achieved the value, the marks are for the working.
- In 'show' questions you must always start with an equation (unless there is a unit conversion prior to the equation, for example). If you do not include the correct equation in a 'show' question, you cannot obtain the marks.
- Marks will be awarded regardless of spelling, as long as meaning is unambiguous. However, the spelling of words such as fusion/fission or refraction/defraction/diffraction needs to be clear.
- Rounding: the significant figure(s) of the final answer can have one figure less or two figures more than the expected answer.

Standard numerical questions

There are techniques you can follow in setting out your answer which will minimise the possibility of simplistic errors. When setting out your working, allow the marker to give marks for correct physics. Arithmetical errors will be penalised but an incorrect formula could be interpreted as *wrong physics* and in these cases the answer will be deemed to be 'wrong' and the marker will stop marking.

Read the question carefully and gather the data from the question on one side of the page. Ensure you are using standard symbols as this will reduce the possibility of selecting an incorrect relationship.

When faced with a standard numerical question, it can help to work through the following steps:
- Check the unit and prefixes. Convert numbers to the correct unit, i.e. mm to m, minutes to seconds, etc.
- Look at the data you have been given and see what you are being asked to do. When you are confident with this, you can select the appropriate relationship from the required list.
- Write down the relationship and then substitute in the figures that you have.
- Carry out the calculation OR rearrange the equation then carry out the calculation (whatever is more comfortable for you).

- Give your final answer as symbol = number unit, such as $F = 156\,\text{N}$. Ensure you have selected the correct unit and that your final answer has the appropriate number of significant figures.

Descriptive, extended-response and open-ended questions

Traditionally, physics candidates are competent in answering numerical questions and dealing with supplied formulae. In contrast, where there has been a requirement to write a paragraph or to explain phenomena using a physics principle, candidates have often struggled. In such questions there may be more than one correct response; marks will be awarded to candidates who understand and can explain the concepts rather than just substituting values into an equation blindly. The following example will illustrate this idea.

Question: A student throws a basketball into a ring from a distance of 9 m as shown in the diagram.

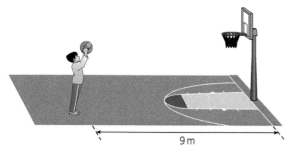

9 m

Figure 1 A student throwing a basketball

He then throws a football of smaller mass with the same velocity and angle. Explain whether or not the ball will go into the ring. (You may ignore air resistance.)

Possible response: The football will go into the ring. The path an object (projectile) follows depends upon the velocity and angle of launch, not the mass of the object. If we can ignore air resistance, the football will follow the same path as the basketball.

An answer like this would get three marks as it explains the physics clearly in a qualitative sense (no numbers or equations). Explanations involving the use of equations are perfectly acceptable, but given there are no indications of distance, height, etc. they may be more difficult.

How you will be assessed in Higher Physics

The course assessment has three components. A multiple-choice paper (worth 25 marks), a written paper (worth 130 marks) and an assignment (worth 20 marks).

The purpose of the question paper is to assess breadth of knowledge from across the sections, depth of understanding and application of this knowledge and understanding to solve problems. It will also test the skills you have acquired during the course.

The assignment requires learners to demonstrate aspects of challenge and application. It will assess the application of skills of scientific investigation/research and using related knowledge, by carrying out a meaningful and appropriately challenging task in physics and communicating findings.

The course assessment is graded A–D and comprises two question papers:
1 The multiple-choice paper has 25 questions each worth a single mark; **25 marks** in total.
2 The written paper will contain restricted-and extended-response questions and will be worth **130 marks**. (This mark of 130 will be scaled down to **95 marks**.)

You will be expected to demonstrate that you have retained knowledge and can apply it to explain observations and phenomena. You will also be expected to demonstrate that you have developed physics skills. Questions assessing both knowledge and understanding and skills may be set in a problem-solving context.

The question paper will give learners an opportunity to:
● integrate and apply physics knowledge and understanding
● apply scientific inquiry skills
● apply scientific analytical thinking skills
● explain the impact of applications on society/the environment.

Approximately 40% of the marks will be assigned to questions related to *Our dynamic Universe*.

Approximately 40% of the marks will be assigned to questions related to *Particles and waves*.

Approximately 20% of the marks will be assigned to questions related to *Electricity*.

A data booklet or sheet containing relevant data and formulae will be provided.

The majority of marks will be awarded for applying knowledge and understanding. The other marks will be awarded for applying scientific inquiry and analytical thinking skills. In addition, around two-thirds of the total marks will be straightforward, while one-third will be at a more demanding level.

The question papers will be given in two sittings. The multiple-choice paper will be of **45 minutes**, duration and the written paper will be of **2 hours and 15 minutes**, duration. This will be under standard examination arrangements.

The question papers are set by SQA and conducted under a high degree of supervision and control. Evidence will be submitted to SQA for external marking. All marking will be quality assured by SQA.

Grade description for C

For the award of grade C, students will have demonstrated successful performance in all of the sections of the course. In the course assessment, candidates will generally have demonstrated the ability to:

- retain knowledge and skills over an extended period of time
- integrate the knowledge, understanding and skills of scientific inquiry acquired across component sections
- apply the knowledge, understanding and skills of scientific inquiry in contexts similar to those in the component sections.

Grade description for A

For the award of grade A, students will have demonstrated successful performance in all of the sections of the course. In addition, candidates at grade A will generally have demonstrated the ability to:

- retain an extensive range of knowledge and skills over an extended period of time
- integrate an extensive range of knowledge, understanding and skills of scientific inquiry acquired across component sections
- apply knowledge, understanding and skills of scientific inquiry in contexts less familiar and more complex than in the component sections.

Assignment

In the assignment you will need to develop the skills necessary to undertake research in physics and demonstrate the relevance of physics to everyday life by exploring the science behind a topical issue. The assignment offers opportunities for collaborative and independent learning set within the context of an evaluation of scientific issues. You will need to collect and synthesise information from a number of different sources.

You should be familiar with carrying out experiments correctly. You will need to plan and undertake a practical investigation related to the topical issue. You will need to prepare a scientific communication, presenting the aim, results and conclusions of the practical investigation.

Carrying out your research

You will be given a topic to investigate. It will be related to something you have covered in your studies, but it could be in an unfamiliar context. You will have to research the physics behind the topic.

Internet searching is preferred by most people these days as the various engines can trawl through all the available sites or articles very quickly, but you need to be careful about the validity of some sites. There are many sites which may appear scientifically valid but which are not as impressive or as accurate as they first seem. Check whether the site has a contact address. This is a good sign.

The URL can also give an indication of a website's authority. For example, if the URL ends in ac.uk this means it is an accredited university or college and would generally be reliable. Reliable websites from America can end in:

.edu – education site

.gov – government site.

It is better if you can obtain evidence from a range of sources and this can be used to support your point. Using more than one source may also give you a better perspective on the topic and, if you can be critical in your use of the information, it will provide evidence of the skills the assignment is hoping to assess.

Report writing

When you carry out your assignment you **must** do at least one experiment or a piece of practical work. This is the data you will process during the communication phase of the assignment. You can collect your experimental data in a group but the group must have a maximum number of four students in it. Your report must be your own work.

During the communication phase you will produce a written report that will be sent to SQA to be marked. The assignment is worth **20 marks** (which will then be scaled to **30 marks** for calculating your final grade).

The marks are awarded in the following categories:
- Aim (1 mark)
- Underlying physics (3 marks)
- Data collection and handling (5 marks)
- Graphical presentation (3 marks)
- Uncertainties (2 marks)
- Analysis (1 mark)
- Conclusion (1 mark)
- Evaluation (3 marks)
- Structure (1 mark)

Aim (1 mark)

The report's aim must be able to be investigated. You should also be careful how you word your aim. An aim that has a 'yes' or 'no' answer is not acceptable. For example, 'To establish the relationship between an LED's switch-on voltage and the wavelength of light produced.' is not acceptable (your conclusion could be 'yes' there is' or 'no' there is not'). However, 'To investigate how an LED's wavelength affects its switch-on voltage' would be acceptable.

Underlying physics (3 marks)

The underlying physics is marked in some ways like an open-ended question. A good response will get 3 marks, a reasonable response will get 2 marks and a limited response will get 1 mark.

Data collection and handling (5 marks)

The data collection is broken down into 5 individual marks:

- 1 mark for a summary of the experimental method. This should not be copied but should be a summary.
- 1 mark for having sufficient raw data from your experiment. If you can repeat an experiment several times you must do so or give an explanation as to why you cannot do repeated measurements.
- 1 mark for having your data and any derived values in a correctly formatted table. This includes units and headings.
- 1 mark for having data from a second source. This can be from another experiment or a reference from the internet or a book.
- 1 mark for a reference to show where your data has come from.

Graphical presentation (3 marks)

In the graphical presentation section, the 3 marks are broken down as follows:

- 1 mark for the graph having suitable scales.
- 1 mark for having the correct units and labels.
- 1 mark for having accurately plotted points and a line (if appropriate) of best fit. When you draw your line, be careful to not force the line through the origin and make sure that the line covers the entire range of your data.

Uncertainties (2 marks)

In the uncertainties section, you should have reading uncertainty in all the measurements that you have made. Take care here, as many students measure something straightforward like distance and then forget to include a reading uncertainty. You need to calculate the random uncertainties in all repeated results. Make sure that you include a sample calculation with your uncertainties. You can also comment on the effect of uncertainties on your investigation. For example, if your graph does not pass close to the origin, you may wish to include a few sentences about systematic uncertainties.

Analysis (1 mark)

In the analysis section, you should try to discuss what your results show. You could compare your results with your internet data, for example, and comment on similarities and differences. If you are working out the value of a constant, then compare your value to the accepted value.

Conclusion (1 mark)

Your conclusion should answer your aim and be supported by the data in your report, for example, if all your data shows that there is a linear relationship between two variables you should say so and not just say that as one increases the other also increases. All your data includes any internet data you have gathered.

Evaluation (3 marks)

There are 3 marks available in the evaluation section. A maximum of one mark is awarded for evaluating internet sources of data so make sure you evaluate your experiment and the data from it carefully. When evaluating an internet source make sure you evaluate the data from the source and not the source itself. One mark is awarded for each correct evaluative statement that is supported by an explanation. If you are evaluating an experimental procedure, either state what happened and explain how it affected your results or give an improvement and explain how it would affect your results. If you state that you would make more repeated measurements, then make sure that this would be an improvement. If your random uncertainties are already small, then doing more repeats will not have any effect. Be careful when evaluating internet sources, as you need to justify your comments about them. You do not need to use words like 'reliable' or 'precise' but, if you do, you must do so correctly.

Structure (1 mark)

The last mark is for a clear and concise report.

Eight hours of study has nominally been allocated to the assignment, of which a maximum of 2 hours is allocated for the report stage. To remind you, the assignment is worth **20 marks** which is scaled up to **30 marks**.

Relationships required for Higher Physics

$d = \bar{v}t$

$s = \bar{v}t$

$v = u + at$

$s = ut + \frac{1}{2}at^2$

$v^2 = u^2 + 2as$

$s = \frac{1}{2}(u + v)t$

$W = mg$

$F = ma$

$E_W = Fd$

$E_p = mgh$

$E_k = \frac{1}{2}mv^2$

$P = \dfrac{E}{t}$

$p = mv$

$Ft = mv - mu$

$F = G\dfrac{Mm_2}{r^2}$

$t' = \dfrac{t}{\sqrt{1-\left(\dfrac{v}{c}\right)^2}}$

$l' = l\sqrt{1-\left(\dfrac{v}{c}\right)^2}$

$f_o = f_s\left(\dfrac{v}{v \pm v_s}\right)$

$z = \dfrac{\lambda_o - \lambda_r}{\lambda_r}$

$z = \dfrac{v}{c}$

$v = H_o d$

$E_w = QV$

$E = mc^2$

$E = hf$

$E_k = hf - hf_o$

$E_2 - E_1 = hf$

$T = \dfrac{1}{f}$

$v = f\lambda$

$d \sin\theta = m\lambda$

$n = \dfrac{\sin\theta_1}{\sin\theta_2}$

$\dfrac{\sin\theta_1}{\sin\theta_2} = \dfrac{\lambda_1}{\lambda_2} = \dfrac{v_1}{v_2}$

$\sin\theta_c = \dfrac{1}{n}$

$I = \dfrac{k}{d^2}$

$I = \dfrac{P}{A}$

Path difference $= m\lambda$ or $(m + \frac{1}{2})\lambda$
where m = 0, 1, 2 …

Random uncertainty $= \dfrac{\text{max. value} - \text{min. value}}{\text{number of values}}$

$V_{peak} = \sqrt{2}V_{r.m.s.}$

$I_{peak} = \sqrt{2}I_{r.m.s.}$

$Q = It$

$V = IR$

$P = IV = I^2R = \dfrac{V^2}{R}$

$R_T = R_1 + R_2 + \ldots$

$\dfrac{1}{R_T} = \dfrac{1}{R_1} + \dfrac{1}{R_2} + \ldots$

$E = V + Ir$

$V_1 = \left(\dfrac{R_1}{R_1 + R_2}\right)V_s$

$\dfrac{V_1}{V_2} = \dfrac{R_1}{R_2}$

$C = \dfrac{Q}{V}$

$E = \frac{1}{2}QV = \frac{1}{2}CV^2 = \frac{1}{2}\dfrac{Q^2}{C}$

Section 1 Our dynamic Universe

Chapter 1

Motion

> ### What you should know
>
> ★ The equations of motion for objects moving with constant acceleration in a straight line
> ★ Motion–time graphs for motion with constant acceleration in a straight line
> ★ Displacement–time graphs, velocity–time graphs and acceleration–time graphs and the interrelationship between them
> ★ Graphs for bouncing objects and objects thrown vertically upwards
> ★ All graphs are restricted to constant acceleration in one dimension, inclusive of change of direction

Equations of motion

The simplest motion that a moving object can have is to be moving in a straight line with a constant **velocity**. When an object exhibits this constant velocity, its **acceleration** is zero. Having zero acceleration means the forces acting on the object must be balanced.

A velocity–time graph of this motion would look like Figure 1.1.

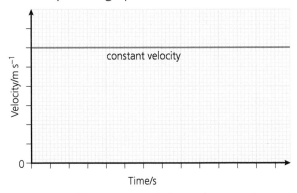

Figure 1.1 A velocity–time graph for an object moving in a straight line with constant velocity

When an object experiences a constant **unbalanced force** it undergoes an acceleration. Its velocity will change by the same amount every second. This change in velocity can be either increasing or decreasing. We will consider increasing velocity first.

1

If an object is moving in a straight line from **rest**, its velocity–time graph will look like Figure 1.2.

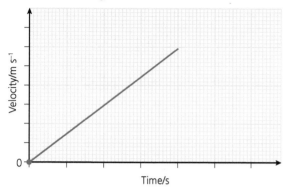

Figure 1.2 A velocity–time graph showing uniform acceleration

If the object is already moving, its velocity–time graph will look like Figure 1.3.

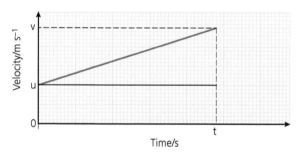

Figure 1.3 The velocity–time graph of an object at velocity u increasing to velocity v after t seconds

The acceleration of the object is equal to the slope of the line – the steeper the slope, the greater the acceleration.

If we use the equation for the slope of the line $m = \frac{y_2 - y_1}{x_2 - x_1}$ we find that:

$$a = \frac{(v - u)}{(t - 0)}$$

This can be rearranged to give:

$$v = u + at$$

The **displacement** of the object, s, is the area under the graph. It can be shown that:

$$s = ut + \frac{1}{2}at^2$$

These two equations can be combined to give a third equation:

$$v^2 = u^2 + 2as$$

It is important to remember that these equations can only be used for an object that is travelling in a straight line with a **constant** acceleration.

Using the equations of motion

The three equations of motion have five variables in them:

- u – initial velocity
- v – final velocity
- a – acceleration
- s – displacement
- t – time taken.

Each equation only has four variables, so if you know three of the five variables it is possible to determine the other two.

Care is needed when using these equations as velocity, acceleration and displacement are all **vectors** and so have a direction. When objects are moving in straight lines, values for these quantities can be either positive or negative. When solving problems with these equations, decide which direction is to be positive and then ensure that anything in the opposite direction has a negative value.

Remember that time can *never* have a negative value. This means that if you calculate a negative value for time, you must have made a mistake and will need to go back and do the calculation again from the beginning.

When attempting problems using the equations of motion, it can be useful to lay out the information from the question in a table or a list and then select the equation you will use.

Example

A ball is brought to rest from a velocity of $12.0\,\text{m s}^{-1}$ in a time of $4.0\,\text{s}$.

a) Calculate the acceleration of the ball.
b) Calculate the displacement of the ball.

Solution

Start this problem by writing down what you know:

$u = 12.0\,\text{m s}^{-1}$
$v = 0\,\text{m s}^{-1}$
$t = 4.0\,\text{s}$
$a = ?$
$s = ?$

a) If we know u, v and t and we want to find a, we use $v = u + at$.
Rearranging this formula gives us $a = \dfrac{v-u}{t}$
Substituting in the values we know from the question:

$a = \dfrac{v-u}{t}$

$a = \dfrac{0-12}{4}$

$a = \dfrac{-12}{4}$

$a = -3.0\,\text{m s}^{-2}$. ⇨

Note that the acceleration is negative. This is because the ball is slowing down. The unbalanced force must be in the opposite direction to its motion.

b) We can now use $s = ut + \frac{1}{2}at^2$ to calculate the displacement.

$s = (12 \times 4) + (\frac{1}{2} \times (-3) \times 4^2)$

$s = 48 + (-24)$

$s = 24\,\text{m}.$

Graphs of motion

Hints & tips ⭐

On a velocity–time graph, the slope/gradient of the line is the value of the acceleration. The area under the graph is the displacement. You can use these two principles to determine the shapes of the acceleration–time graph and the displacement–time graph for the same motion.

If the slope of the line is positive then the acceleration is positive; if the slope of the line is negative then the acceleration is negative. The steeper the line, the greater the acceleration.

If the object is accelerating then the area under the velocity–time graph is increasing by a larger amount each second, therefore the displacement–time graph for this motion will be an ever-increasing curve. If the velocity is constant, the increase in area under the graph is the same each second so the displacement–time graph has a line with a constant slope.

When an object is moving with a constant velocity, its acceleration is zero and its displacement increases at a constant rate. This is represented in the graphs in Figure 1.4.

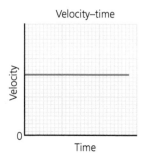

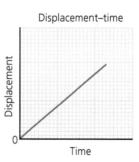

Figure 1.4 The graphs of motion for an object moving with a constant velocity

An object moving with a constant positive acceleration increases its velocity at a constant rate. Its displacement increases by a greater amount each second. These are difficult concepts and it can take time and many worked examples for the meaning to become clear. These ideas are represented by the graphs in Figure 1.5.

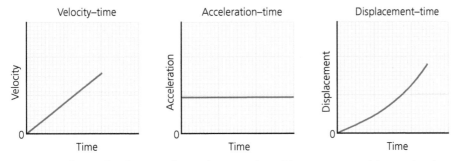

Figure 1.5 The graphs of motion for an object moving with a constant positive acceleration

For an object moving with a constant negative acceleration, its velocity will decrease at a constant rate but its displacement will increase by smaller increments each second. This is represented by the graphs in Figure 1.6.

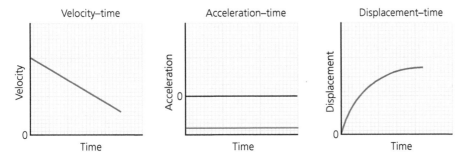

Figure 1.6 The graphs of motion for an object moving with constant negative acceleration

It is important that you know these three groups of graphs and understand how each graph relates to the others in its group.

There are two other velocity–time graphs that you should know and understand. The first concerns an object thrown vertically upwards and then coming back down.

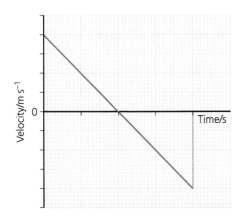

Figure 1.7 The velocity–time graph for an object thrown vertically upwards and then caught

The gradient of this graph is a constant negative value. This is because the acceleration of the object is always downwards and in the opposite direction to the one it was thrown in. The total area under the graph is zero. This is because the final displacement of the object is zero. Its upwards displacement is the same in size but opposite in direction to its downwards displacement. It has travelled 2·4 m upwards then 2·4 m downwards, for example.

The second graph that should be remembered is that showing the motion of a bouncing ball (Figure 1.8).

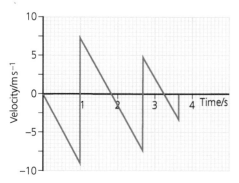

Figure 1.8 The velocity–time graph of a bouncing object

The slope of the line for the majority of the graph is a constant negative value. The sections where the graph is nearly a vertical line represent the motion of the ball when it is in contact with the ground and rebounding. The negative slope is due to the ball's downwards acceleration regardless of its upwards or downwards motion. The area under the graph reduces with each bounce as the total displacement reduces. Each time the ball bounces, its rebound height is lower.

Key points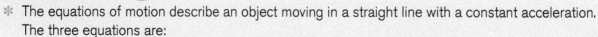

* The equations of motion describe an object moving in a straight line with a constant acceleration. The three equations are:

$$v = u + at$$
$$s = ut + \frac{1}{2}at^2$$
$$v^2 = u^2 + 2as$$

  where: u = initial velocity; v = final velocity; a = acceleration; t = time; s = displacement.
* Velocity, acceleration and displacement are all vectors and so require a direction.
* When objects are moving in a straight line, make one direction (say to the right) positive and then anything in the opposite direction is negative. Later on, when dealing with projectiles and forces, you will need to think more deeply about this.
* Time can never have a negative value. If you calculate a negative time, you must have made a mistake so you should redo the calculation from the start.
* The motion of an object moving in a straight line with a constant acceleration can be analysed using velocity–time, acceleration–time and displacement–time graphs.
* There is a relationship between the shapes of the velocity–time, acceleration–time and displacement–time graphs for a particular type of motion.
* You should be able to draw the three graphs for a constant velocity, a constant acceleration and a constant deceleration.
* You will need to be able to draw graphs for an object thrown vertically upwards and a bouncing object.
* It is important that you can interpret all the graphs listed above.

Key words

Acceleration – the rate of change of velocity
Balanced forces – equivalent to having no force acting on an object; the forces cancel each other out
Constant – not changing; uniform
Displacement – how far an object is from a point in a particular direction
Rest – not moving; stationary
Scalar – a quantity that has magnitude (size) and a unit
Unbalanced forces – forces that cause an object to accelerate
Vector – a quantity that has magnitude (size), a unit and a direction
Velocity – the rate of change of displacement

Questions

1 An object is moving with a constant velocity. What can be said about the forces acting on it?
2 An object accelerates from $2.0\,\mathrm{m\,s^{-1}}$ to $8.0\,\mathrm{m\,s^{-1}}$ at $1.5\,\mathrm{m\,s^{-2}}$. Calculate the object's displacement during its motion.
3 A car brakes from $12\,\mathrm{m\,s^{-1}}$ to rest with a deceleration of $4.0\,\mathrm{m\,s^{-2}}$. Calculate the time it takes for the car to come to rest.
4 Figure 1.9 shows how the velocity of an object varies with time. Sketch the corresponding acceleration–time and displacement–time graphs for this object. (Numerical values are not required.)

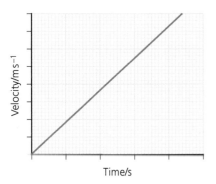

Figure 1.9 Graph of motion

5 Figure 1.10 shows how the displacement of an object varies with time. Sketch the corresponding velocity–time and acceleration–time graphs for this object. (Numerical values are not required.)

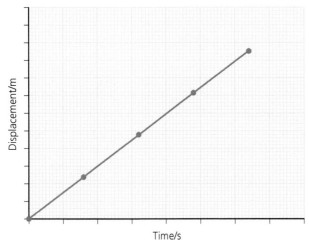

Figure 1.10 Graph of motion

⇨

6 Figure 1.11 shows how the velocity of a bouncing ball varies with time.

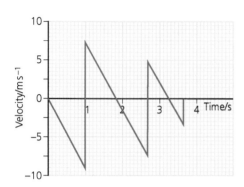

Figure 1.11 Graph of motion of a bouncing ball

a) Why is the gradient of the line constant after each bounce?
b) When the time is 2·0 s, in which direction is the ball travelling?
c) The graph shows that the ball first hits the ground 1·0 s after it is dropped. Calculate the height from which the ball is dropped.

*7 A golfer hits a ball from point P. The ball leaves the club with a velocity, v, at an angle of θ to the horizontal. The ball travels through the air and lands at point R. Midway between P and R there is a tree of height 10·0 m.

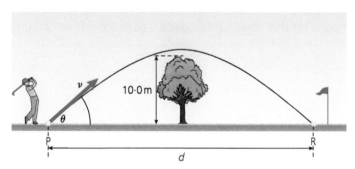

Figure 1.12 Golfer hitting a ball

a) The horizontal and vertical components of the ball's velocity during its flight are shown.

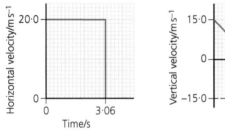 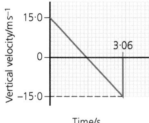

Figure 1.13 Graphs showing horizontal and vertical components of velocity

The effects of air resistance can be ignored. Calculate:
   (i) the horizontal distance, d
   (ii) the maximum height of the ball above the ground.
b) When the effects of air resistance are **not** ignored, the golf ball follows a different path. Is the ball more or less likely to hit the tree? You must justify your answer.

Chapter 2

Forces, energy and power

What you should know

* ★ Balanced and unbalanced forces; the effects of friction; terminal velocity
* ★ Forces acting in one plane only
* ★ Analysis of motion using Newton's First and Second Laws; frictional force as a negative vector quantity (no reference to static and dynamic friction)
* ★ Tension as a pulling force exerted by a string or cable on another object
* ★ Velocity–time graph of a falling object when air resistance is taken into account, including the effect of changing the surface area of the falling object
* ★ Resolving a force into two perpendicular components
* ★ Forces acting at an angle to the direction of movement
* ★ Resolving the weight of an object on a slope into a component acting down the slope and a component acting normal to the slope
* ★ Systems of balanced forces with forces acting in two dimensions
* ★ Work done, potential energy, kinetic energy and power in familiar and unfamiliar situations
* ★ Conservation of energy

Hints & tips ★

In the 'what you should know' box (left), the following phrase appears: 'no reference to static and dynamic friction'. This is not something you will need to worry about. It is merely there to tell teachers that they do not need to make a difference between the type of friction acting on a moving object and the type of friction acting on a stationary object.

Forces and their effects on the motion of an object will have been introduced at National 4 and 5. You will already be aware that a force can cause an object to change its velocity, direction or shape. When a force is applied to a stationary trolley, it will accelerate. It will increase its velocity for as long as that force is applied or until other forces are applied. Most situations involve more than one force being applied and difficulty arises when the combination of forces has to be determined.

When an object's velocity is not changing, this can be a result of two conditions:

1 There are no forces acting on the object.
2 The forces acting on the object cancel each other out (they are balanced).

Hints & tips ★

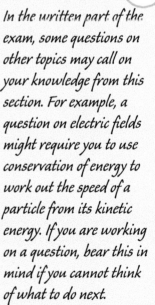

In the written part of the exam, some questions on other topics may call on your knowledge from this section. For example, a question on electric fields might require you to use conservation of energy to work out the speed of a particle from its kinetic energy. If you are working on a question, bear this in mind if you cannot think of what to do next.

Examples of balanced forces

1 A boat floating while stationary. It is not moving or changing its velocity therefore forces are balanced. The weight of the boat downwards is 'balanced' by the buoyancy force acting upwards.

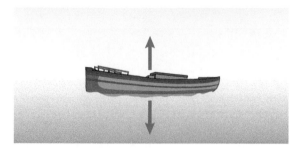

Figure 2.1 The forces on this boat are balanced – the weight downwards is balanced by the buoyancy force upwards.

2   A mass hanging on a spring. The weight of the mass acting downwards is balanced by the force of the spring acting upwards.

Figure 2.2 The weight of the mass is balanced by the upwards force of the spring.

3   A car driving at a steady 15 m s⁻¹. Its velocity is not changing therefore forces are balanced. The force from the engine driving it forwards is balanced by the friction from the road and air (drag).

Figure 2.3 The forward force on this car is balanced by the backwards friction force.

4   A raindrop falling to the ground. A raindrop falls at a steady velocity. The weight of the raindrop acting downwards is balanced by the drag acting in the opposite direction. It falls at a steady velocity.

Figure 2.4 The weight of the raindrop is balanced by the upwards drag force.

The weight of the object suspended from the crane in Figure 2.5 is balanced by the force from the cable holding it up. The cable 'transfers' this force to the jib of the crane and the jib must be able to withstand this force. The cable is said to be in **tension**. The cable is being pulled and it can apply a force to an object that it is in contact with.

Figure 2.5 A crane amidst skyscrapers

Friction

Friction is a force that opposes the motion of moving objects. It acts in almost every situation we can imagine. It is caused by the interaction of one surface with another; rubber soles on pavements, tyres on roads, hinges on doors turning, etc. In physics we will consider friction acting on moving objects.

A force acting on a stationary block causes it to accelerate. As it moves, the block is in contact with the surface it rests upon and a frictional force acts to slow it down.

Friction is a force acting against the direction of motion. This allows us to treat it as a vector acting in the other direction; it has a negative effect.

Figure 2.6 The friction acts on the movement of the block to slow it down.

> ### Example
>
> A car accelerates from rest. It has a mass of 1250 kg and the engine produces a force of 880 N.
>
> Its initial acceleration, a, is $\frac{F}{m} = \frac{880}{1250} = 0\cdot7\,\text{m}\,\text{s}^{-2}$.
>
> After 3 seconds the frictional forces acting on the car are 410 N. At this point the resultant force driving the car is $880 - 410 = 470$ N, giving an acceleration (a) of $\frac{F}{m} = \frac{470}{1250} = 0\cdot38\,\text{m}\,\text{s}^{-2}$.

The frictional force is *not* slowing the car down; it is reducing its acceleration. Ultimately, at some speed the frictional force will equal the driving force and the forces will be balanced. It will remain at that velocity and *not* accelerate. This is why we have to keep our foot on the accelerator to keep a car moving at 40 mph, for example.

This combination of forces is what limits the maximum velocity a car or moving object can reach. For an object to have a great maximum velocity it must have an engine that produces a large force and a shape that minimises the frictional forces. This rule holds true for all vehicles, whether trains, aeroplanes or boats.

Terminal velocity

The term **terminal velocity** applies to objects that have reached their maximum velocity. An object released and allowed to fall will have an initial acceleration of $9\cdot8\,\text{m}\,\text{s}^{-2}$. As the object increases its velocity, the frictional forces (or drag) increase. This drag acts in the opposite direction to its weight. This combination of weight (downwards) and drag (upwards) results in a smaller force downwards, leading to a reduced acceleration. Ultimately the drag force will reach the same value as the weight. At this point the forces on the object are balanced and will result in an overall downwards force of zero. The object will not go any faster and we say that it has reached its terminal velocity.

The graph in Figure 2.7 shows the velocity of a falling object with frictional effects.

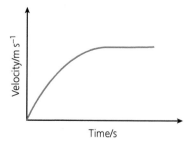

Figure 2.7 Graph to show the velocity of a falling object

The **magnitude** of the terminal velocity of a falling object depends upon two factors: weight and friction (drag). If the drag of an object is increased, its terminal velocity will decrease. Skydivers can increase and decrease their terminal velocity by changing their body shape. This ability by skydivers to control their velocity allows them to catch up with each other and create spectacular formations in the air. Animals can also change their shapes in order to increase velocity.

Figure 2.8 Both skydivers and animals such as this red kite can control their terminal velocity by altering their shape in the air.

When a skydiver releases the parachute, drag suddenly increases and the terminal velocity reduces dramatically.

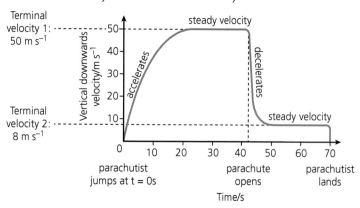

Figure 2.9 This graph shows the change in velocity at the point when a skydiver's parachute is opened.

In general, heavy objects will have a greater terminal velocity than light objects.

An object will accelerate when a force is applied. Falling objects are acted upon by gravitational attraction. We can also apply a force to an object with a cable or spring.

Resolving forces

A force is a **vector** quantity; it has a size (magnitude) and a direction. Unless the two vectors acting on an object are in the same plane and direction (as they have been in our previous examples), we cannot simply add or subtract their magnitude. However, forces can act in any direction. We need to use the fact that a vector can be resolved into **perpendicular** (at right angles) **components**; mainly horizontal and vertical.

The 65 N vector shown in Figure 2.10 can be resolved into horizontal and vertical components as given below:

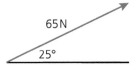

H = 65 cos 25° = 58·9 = 59 N

V = 65 sin 25° = 27·47 = 27 N.

Figure 2.10 Vector

This resolution of vectors can be applied to other examples where a force is acting on a body in a direction other than the direction of travel.

Examples

1 Calculate the component of the force acting in the direction of motion and hence the total force being applied.

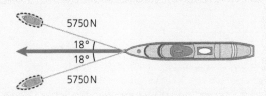

Figure 2.11 Two tugboats pulling a ship

Solution

Force to the left from one cable:
F = 5750 × cos 18° = 5468·6 N.
There are two cables, therefore the total force is:
2 × 5468·6 = 10937·2 = 10900 N.

The components of the tension forces acting at 90° to the ship are equal and act in the opposite direction, thereby 'cancelling' each other out.

2 A person drags a box across a floor.

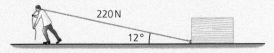

Figure 2.12 A person dragging a box along the floor

Calculate the force pulling the block in the horizontal direction.

Solution

F = 220 × cos 12° = 215·2 N = 215 N.

It is this force that will have to overcome the friction on the block in order for it to move.

The same method can be applied to an object moving down a slope.

An object on a horizontal surface will not move as the weight is acting in a downwards direction and the only direction in which the block can move is horizontally. If the surface is gradually elevated at one end, the slope will eventually be great enough so that the object will move down the slope.

The force moving the object is the component of the object's weight acting down the slope. In these examples the weight of the object is resolved into a component *onto* the slope and a component *down* the slope.

The force acting down the slope is calculated using $F_d = mg \sin \theta$.

The force acting onto the slope is calculated using $F_o = mg \cos \theta$.

The force down the slope increases with the angle of the slope. It varies directly as the sine of the angle of the slope.

Hints & tips

The equations that give the component of weight down the slope and the component of weight acting perpendicular to the slope are not on the data sheet included with the exam. This means that you will either have to remember them or work them out if you need to use them.

The component of weight down the slope (F = mg sin θ) is used frequently in the exam so it is important to know this equation.

Examples

Example of a block on a frictionless slope

Figure 2.13 A block on a slope

Calculate the component of the weight acting down the slope and the acceleration of the block.

Solution

$F = mg \sin \theta = 4.5 \times 9.8 \times \sin 22° = 16.52 = 17\,N$

Acceleration $= \dfrac{F}{m} = \dfrac{17}{4.5} = 3.77 = 3.8\,m\,s^{-2}$

Example of a block on a slope with friction being considered

A 2·5 kg block is released on a slope at 28° to the horizontal and the friction between the block and slope is 3·6 N

Calculate:

a) the component of weight down the slope
b) the unbalanced force acting on the block
c) its resulting acceleration.

⇨
Solution

a) $F = mg \sin \theta = 2 \cdot 5 \times 9 \cdot 8 \times \sin 28° = 11 \cdot 5 = 12 \, \text{N}$
b) Unbalanced force $= 12 - 3 \cdot 6 = 8 \cdot 4 \, \text{N}$
c) Acceleration $= \dfrac{F}{m} = \dfrac{8 \cdot 4}{2 \cdot 5} = 3 \cdot 36 = 3 \cdot 4 \, \text{m s}^{-2}$

Effect of friction when an object is moving up the slope

Using the same slope and block as in the example above, consider what would happen if the block was initially pushed up the slope.

The force acting down the slope is the component of the weight acting down *and* the frictional force (as frictional forces act against the direction of motion).

Solution

Unbalanced force $= 12 \, \text{N} + 3 \cdot 6 \, \text{N} = 15 \cdot 6 \, \text{N}$
Acceleration of the block, $a = \dfrac{F}{m} = \dfrac{15 \cdot 6}{2 \cdot 5} = 6 \cdot 24 = 6 \cdot 2 \, \text{m s}^{-2}$
The acceleration of the block is greater going up the slope than coming down. This can seem unusual as it is the same block and slope. However, the direction of motion has changed and this affects the overall conditions.

Work, power and energy

One of the fundamental principles in physics is the **Law of Conservation of Energy**. It states that in a closed system, energy cannot be created or destroyed. When a system such as that involving a trolley rolling down a slope is analysed, it can be shown that the total energy before the trolley rolls down is the same as the total energy once the trolley has reached the bottom. The energy has been transformed from certain forms at the beginning of the experiment to other forms at the end. No energy has been lost, only converted to other forms.

The **potential energy** (E_p) the trolley has by virtue of being at the top of the slope has been transformed to **kinetic energy** (E_k), or movement energy, as it rolls down the slope. Some sound energy has been caused by the wheels; there may also be some heat energy generated due to friction on the surface and in the bearings.

You should be able to solve problems of this type by using various energy relationships to find a particular answer.

There will be questions in your course assessment which will require you to provide a descriptive explanation of a phenomenon to gain marks. The following gives an example of such a question.

Figure 2.14 No energy is lost as the trolley rolls down the slope.

Hints & tips

*Occasionally an exam question may ask you to **show** rather than **calculate** a value. This is usually because you need to use the value in a later part of the question and if you did not have the value you would be stuck. Be careful when you do this type of question as the marks are for the working. You need to show all the steps you take or you will lose marks.*

Example

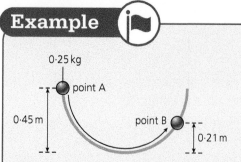

Figure 2.15 Motion of a ball in a curved U-shaped track

Describe and explain the motion of the ball as it moves from point A to point B in Figure 2.15.

Solution

The ball has potential energy at point A. It is released and its potential energy is gradually converted to kinetic energy. As it rolls along the track, the ball has to overcome friction and some of its kinetic energy is converted to heat. This reduces its velocity. It then travels up the opposite slope but as it has less kinetic energy now, it will not reach the same height as the height from which it was released.

Example

Experiment to measure acceleration

You need to be able to describe an experiment to measure the acceleration of an object moving down a slope.

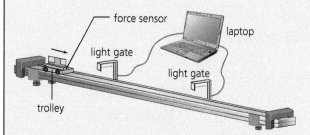

Figure 2.16 An experiment to measure acceleration

The trolley is released and it moves down the ramp, breaking the beams across the light gates as it travels.

The card on the trolley is measured and the length noted (d).

The card passes through gate 1 and the time (t_1) to pass is recorded. The trolley continues and passes through gate 2, where the time (t_2) to pass is also recorded. The computer also records the time taken for the trolley to travel from gate 1 to gate 2.

The velocity of the trolley through gate 1 is calculated using $v_1 = \dfrac{d}{t_1}$.

The velocity of the trolley through gate 2 is calculated using $v_2 = \dfrac{d}{t_2}$. The time to pass between gates is t.

The acceleration of the trolley can be calculated using $a = \dfrac{v_2 - v_1}{t}$.

Key points !

* Balanced forces acting on an object are equivalent to no force acting on the object.
* When balanced forces act on an object, the motion of an object does not change.
* This leads to Newton's First Law of Motion – an object will remain at rest or continue to move in a straight line at a constant speed unless an unbalanced force acts upon it.
* An unbalanced force will produce an acceleration on an object.
* Acceleration and velocity are vector quantities. When an object's direction of travel changes, its velocity changes even if its speed stays the same. When an object's direction is changing, it is accelerating.
* Friction is a force that always acts in the opposite direction to the direction in which an object is travelling.
* The faster an object moves, the greater the friction force acting on it.
* When the friction force is equal in size but opposite in direction to the other force or forces on an object, the forces are balanced and so the object stops accelerating. This is the terminal velocity of the object.
* Newton's Second Law of Motion states that the acceleration on an object is proportional to the unbalanced force and inversely proportional to the mass of the object. We usually write this as $F = ma$. Be careful with this equation; F is the *unbalanced* force.
* You should have an understanding of how to use Newton's laws to analyse motion and be able to apply them to objects that are stationary, including situations where objects are being towed or suspended with cables or ropes.
* You should be able to draw and compare graphs of the motion of an object when friction is ignored and when friction is taken into account.
* You should understand how to find the perpendicular components of force vectors.
* You should be able to describe what happens to an object when a force acts at right angles to its direction of travel.
* You should understand the forces acting on an object on a slope.
* The component of weight of an object acting down the slope $= mg \sin \theta$.
* The component of weight of an object acting normal to the slope $= mg \cos \theta$.
* Work done is the energy transferred during a process. Work done = force × distance. Note that the force must be in the same direction as the distance moved.
* You should be able to solve problems where energy is converted from one form into another.
* Energy cannot be created or destroyed, only changed from one form into another.

Key words

Component – part of something
Conservation – the total amount of a quantity stays the same
Friction – the force that opposes the motion of all objects
Magnitude – the size of something
Perpendicular – at right angles
Terminal velocity – the velocity of an object when the driving force acting on it is balanced by the friction force
Work done – a measure of the energy transferred or converted during a process

Questions ?

1 Indicate the forces acting in the plane of motion in the following examples:
   a) a fish swimming towards some reeds
   b) a ball moving along the surface of a pitch
   c) a skier sliding down a snowy slope
   d) a rocket that has just been launched.

2 Calculate the horizontal and vertical components of the following vectors:
   a) 125 m s^{-1} at 12° to the horizontal
   b) 70·0 N at 65° to the horizontal
   c) 650 N at 40° to the horizontal.

3 A block of 12 kg is raised to a height of 2·5 m.
   a) Calculate its gain in potential energy.
   It is then released and allowed to fall to the ground.
   b) Assuming all the potential energy is converted to kinetic energy, calculate its velocity as it hits the ground.
   c) Some students repeat the above experiment but set up light gates to measure the actual velocity of the block as it hits the ground. They calculate the velocity to be 6·1 m s^{-1}. The experiment is repeated and a similar result is obtained.
      (i) Calculate the difference between the actual kinetic energy of the block and the value used in part b).
      (ii) Account for this difference in kinetic energy.

4 A pile is driven into the ground by means of a large mass of 450 kg being raised to a height of 1·2 m and then released.
   a) Calculate the velocity of the mass as it just hits the pile (assuming no losses due to air resistance).
   The impact of the mass with the pile transfers 3500 J of energy to the 1200 kg pile. The impact drives the pile 0·14 m into the ground.
   b) Give a value for the force of friction between the pile and the Earth. (Ignore potential energy of pile in this example.)

5 A toy car of mass 0·15 kg has a motor that can be 'charged' prior to being released. It receives 0·35 J from the charging unit. It is released and moves off.
   a) Assuming all the energy is converted to kinetic energy, calculate its velocity on release.
   b) The constant frictional force between the car and the floor of the lab is 0·09 N. Calculate how far it will travel before stopping.
   c) The same car is released on an upwardly pointing slope, which also has a constant frictional force of 0·09 N, where it travels a distance of 1·4 m. What height did the car reach?

6 A box of mass 0·24 kg is moving at a constant speed of 0·16 m s^{-1} down a slope. The angle the slope makes with the horizontal is 30°.
   a) (i) Calculate the kinetic energy of the box.
      (ii) Show that the magnitude of the frictional force acting up the slope is 1·2 N.
   b) The box travels 4·2 m down the slope at this speed. Calculate the work done against friction when travelling this distance.

7 A trolley of mass 2·0 kg is at the top of a frictionless slope which is at an angle of 15° to the horizontal. The height of the slope is 1·6 m. The trolley is allowed to run down the slope.
   a) Calculate:
      (i) the gravitational potential energy of the trolley at the top of the slope
      (ii) the speed of the trolley at the bottom of the slope.
   b) The angle of the slope is increased to 25° without changing the height from which the trolley is released. Will the speed of the trolley be greater than, less than or the same as in part a)? Justify your answer.
   c) The mass of the trolley is increased to 4·0 kg and it is again allowed to run down the frictionless slope. Will the speed of the trolley be greater than, less than or the same as in part a)? Justify your answer.

8 A block of mass 1·2 kg is dragged along a ramp at a constant speed by a force of 20 N. The ramp makes an angle to the horizontal of 30° and is 6·0 m long. Calculate:
   a) the work done against friction in moving the block along the slope
   b) the potential energy gained by the block.

9 A ball is rolled up a slope. As it goes up the slope it slows down, its velocity reaches zero and then it starts to roll back down the slope. When it goes up the slope, the magnitude of acceleration is measured to be 1·6 m s^{-2} but as it rolls down the slope, the acceleration is found to be 1·2 m s^{-2}. Account for the difference.

10 A drone is hovering over a piece of ground. A parcel is suspended from a cable beneath the drone as shown.
   a) The mass of the drone and parcel is 9.75kg. Show that the drone produces a lift force of 95.6N.
   b) The drone now releases the parcel which has a mass of 2.6kg. Describe the vertical motion of the drone immediately after releasing the parcel. Justify your answer in terms of the forces acting on the drone.

Figure 2.17 Parcel suspended by a cable

*11 A car is travelling at a constant speed of 15 m s^{-1} along a straight, level road. It passes a motorcycle which is stationary at the roadside.

Figure 2.18 Car and motorbike

At the instant the car passes, the motorcycle starts to move in the same direction as the car. The graph shows the motion of each vehicle from the instant the car passes the motorcycle.
   a) Show that the initial acceleration of the motorcycle is 5·0 m s^{-2}.
   b) Calculate the distance between the car and the motorcycle at 4·0 s.
   c) The total mass of the motorcycle and rider is 290 kg. At a time of 2·0 s, the driving force on the motorcycle is 1800 N.
      (i) Calculate the frictional force acting on the motorcycle at this time.
      (ii) Explain why the driving force must be increased with time to maintain a constant acceleration.

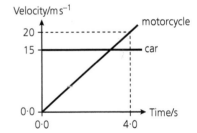

Figure 2.19 Line graph

Chapter 3

Collisions, momentum and energy

What you should know

★ Conservation of momentum in one dimension and in cases where the objects may move in opposite directions
★ Kinetic energy in elastic and inelastic collisions
★ Explosions and Newton's Third Law of Motion
★ Conservation of momentum in explosions in one dimension only
★ Force–time graphs during contact of colliding objects
★ Impulse found from the area under a force–time graph
★ Equivalence of change in momentum and impulse

Momentum

In this section we will be looking at how objects interact with each other when they collide or when they move apart in an explosion.

When looking at these situations, it is useful to have a measurable quantity to help us understand the interactions. The quantity that we use is called **momentum**.

The momentum of an object, p, is simply its mass multiplied by its velocity:

$$p = m \times v$$

Momentum is a vector quantity. This means that when you analyse collisions and explosions, the direction that objects travel in is important. In this course all the collisions and explosions will take place in a single straight line. When you are working on a collision or explosion problem, choose one direction (say, to the right) and make it positive. Anything moving in the opposite direction will have a negative value.

The unit for momentum is $kg\,m\,s^{-1}$. This originates from momentum being equal to mass multiplied by velocity.

It is useful to realise that the momentum of any stationary object is zero.

Collisions

When two objects collide, they affect each other's motion. In physics terms what happens is that one object transfers some momentum to the other. In the real world, collisions are very complex.

Figure 3.1 A spectacular collision

In this chapter we will consider simple collisions under specified conditions. As discussed previously, the collisions we study will all take place in a single straight line. In addition, we will assume that the objects are moving over frictionless surfaces.

During any collision, if there are no external forces (like friction), the total momentum of all the objects before the collision is equal to the total momentum of all the objects after the collision. This does not mean that the momentum of an object does not change; it means that if one object loses momentum, this will be gained by the other object. In other words, the total momentum does not change but the momentum of individual objects can change during the collision. This principle is called the **Law of Conservation of Linear Momentum**. It is usually stated in these terms:

In the absence of external forces, the total momentum before the collision is equal to the total momentum after the collision.

Let us consider an example of a simple collision.

Figure 3.2 Before a collision   **Figure 3.3** After a collision

We can work out the momentum before and after the collision.

> mass of ball A, $m_A = 4.0\,\text{kg}$
>
> mass of ball B, $m_B = 2.0\,\text{kg}$

Before the collision:

> velocity of ball A, $u_A = 2.5\,\text{m s}^{-1}$
>
> velocity of ball B, $u_B = 0\,\text{m s}^{-1}$
>
> total momentum before the collision $= m_A u_A + m_B u_B$
>
> $= (4.0 \times 2.5) + (2.0 \times 0) = 10 + 0 = 10\,\text{kg m s}^{-1}$

After the collision:

velocity of ball A, $v_A = 1.0\,\text{m}\,\text{s}^{-1}$

velocity of ball B, $v_B = 3.0\,\text{m}\,\text{s}^{-1}$

total momentum after the collision $= m_A v_A + m_B v_B$

$= (4.0 \times 1.0) + (2.0 \times 3.0) = 4 + 6 = 10\,\text{kg}\,\text{m}\,\text{s}^{-1}$

We can see that the momentum before the collision is equal to the momentum after the collision. We say that momentum is **conserved**.

We can write this as an equation:

$$m_A u_A + m_B u_B = m_A v_A + m_B v_B$$

Now let us consider a collision between the same two balls as before but instead of one being stationary, we will see what happens if the second ball is also moving towards the first.

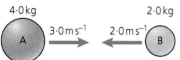

Figure 3.4 Before the collision   **Figure 3.5** After the collision

We can calculate the momentum before and after the collision.

mass of ball A, $m_A = 4.0\,\text{kg}$

mass of ball B, $m_B = 2.0\,\text{kg}$

Before the collision:

velocity of ball A, $u_A = 3.0\,\text{m}\,\text{s}^{-1}$

velocity of ball B, $u_B = -2.0\,\text{m}\,\text{s}^{-1}$ (Remember that the velocity is
negative because ball B is travelling
in the opposite direction to A.)

total momentum before the collision $= m_A u_A + m_B u_B$

$= (4.0 \times 3.0) + (2.0 \times (-2.0)) = 12 - 4 = 8\,\text{kg}\,\text{m}\,\text{s}^{-1}$

After the collision:

velocity of ball A, $v_A = 0.5\,\text{m}\,\text{s}^{-1}$

velocity of ball B, $v_B = 3.0\,\text{m}\,\text{s}^{-1}$

total momentum after the collision $= m_A v_A + m_B v_B$

$= (4.0 \times 0.5) + (2.0 \times 3.0) = 2 + 6 = 8\,\text{kg}\,\text{m}\,\text{s}^{-1}$

Again, it can be seen that the total momentum before the collision equals the total momentum after the collision.

We can use what we have just learned to calculate what happens during a collision.

Consider the following collision:

A 6·0 kg trolley travelling at 2·0 m s⁻¹ collides with a stationary 4·0 kg trolley. After the collision, the two trolleys stick together and move off together.

Figure 3.6 Before and after the collision of two trolleys

We can use the equation for the conservation of linear momentum to calculate the velocity of the trolleys after the collision.

mass of trolley A, m_A = 6·0 kg

mass of trolley B, m_B = 4·0 kg

Before the collision:

velocity of trolley A, u_A = 2·0 m s⁻¹

velocity of trolley B, u_B = 0 m s⁻¹

We do not know the velocity of either trolley after the collision, but as they are moving together, we know that it must be the same.

Velocity of trolley A, v_A = velocity of trolley B, v_B = v

We can use the equation for the conservation of linear momentum and substitute in our values:

$$m_A u_A + m_B u_B = m_A v_A + m_B v_B$$
$$(6 \times 2) + (4 \times 0) = (6v) + (4v)$$
$$12 + 0 = 10v$$
$$v = 12 \div 10$$
$$v = 1·2 \text{ m s}^{-1}$$

The two trolleys move to the right at 1·2 m s⁻¹.

Hints & tips

When carrying out calculations for collisions and explosions it can be useful to draw a table of the six variables and insert the values.

m_A =	u_A =	v_A =
m_B =	u_B =	v_B =

Now the data can be inserted into the equation and the problem solved:

$$m_A u_A + m_B u_B = m_A v_A + m_B v_B$$

Types of collision

There are two types of collision, **elastic** and **inelastic**. We define the two types of collision the following way:

- In an elastic collision, kinetic energy is conserved.
- In an inelastic collision, kinetic energy is not conserved (it decreases).

It is important to note that total energy during a collision is always conserved, but in an inelastic collision, kinetic energy gets converted into other forms of energy (mainly heat).

To determine whether a collision is elastic or inelastic it is necessary to calculate the total kinetic energy before the collision, the total kinetic energy after the collision and then compare the two values.

The total kinetic energy before the collision $= \frac{1}{2}m_A u_A^2 + \frac{1}{2}m_B u_B^2$

The total kinetic energy after the collision $= \frac{1}{2}m_A v_A^2 + \frac{1}{2}m_B v_B^2$

If the total kinetic energy before equals the total kinetic energy after, we can say that the collision is elastic.

If the total kinetic energy before is greater than the total kinetic energy after, we know that the collision is inelastic.

Almost all collisions in the real world are inelastic. It is only in situations like gas particles colliding (at the atomic level) that collisions are perfectly elastic.

Explosions

In an **explosion** two objects move apart. Explosions do not have to be explosive in the everyday sense of the word. When two trolleys are pushed apart by a spring we call this an explosion.

Figure 3.7 An example of a simple explosion

In explosions, momentum is conserved just as it is in collisions. This means that in the absence of external forces the total momentum before the explosion is equal to the total momentum after the explosion.

Often the objects are not moving prior to an explosion. In such cases, the total momentum before the explosion is zero. This means that the total momentum after the explosion must also be zero. This does not mean that after the explosion the objects will *not* be moving; momentum is a vector quantity and if the momentum of the object moving to the right is the same as the momentum of the object moving to the left, the total momentum will be zero. The two values for the 'momentums' are the same in size but opposite in direction, so one will be positive and the other negative and hence they cancel each other out.

When we analyse explosions we use the same equation as for collisions, that is:

$$m_A u_A + m_B u_B = m_A v_A + m_B v_B$$

You need to consider the directions carefully when using this equation.

During an explosion the kinetic energy of the system increases. We know that total energy must always remain constant, so the kinetic energy must have come from some stored energy, usually from a spring or a chemical.

Newton's Third Law and collisions and explosions

Newton's Third Law of Motion states that for every action there is an equal and opposite reaction.

In terms of collisions and explosions this means that the force one object exerts on the other is equal in size but opposite in direction to the force that the second object exerts on the first.

You should also remember that in collisions and explosions, objects exchange momentum. This means that the momentum that one object gains is equal in size to the momentum lost by the other object. The change in momentum of object A is equal in size but opposite in direction to the change in momentum of object B.

Impulse and change in momentum

When considering the forces applicable during collisions it is useful to use the quantity, **impulse**. The impulse that an object receives during a collision or explosion is the average force multiplied by the time of contact. Impulse is a vector quantity and is measured in N s.

We know that $F = ma$ and $a = \frac{v-u}{t}$.

This leads to $F = m\left(\frac{v-u}{t}\right)$

$$F = \frac{mv - mu}{t}$$

$$Ft = mv - mu$$

The term Ft is the impulse on an object and $mv - mu$ is the change in momentum of an object. We can see that the impulse is equal to the change in momentum of an object. As a result, an alternative unit for impulse is $kg\, m\, s^{-1}$.

We have already learned that the change in momentum of one object is equal in size but opposite in direction to the change in momentum of the other object. This means that the impulse on one object is equal in size but opposite in direction to the impulse on the other object. This also fits in with Newton's Third Law as the force on one object is equal in size but opposite in direction to the force on the other object.

Example

You need to be very careful of the sign convention during collisions. If an object rebounds then its new velocity has the opposite sign to its original velocity.

This is a very common mistake in exam answers so care is needed when doing this type of calculation. An example is shown below.

A ball of mass $4.0\,kg$ is travelling to the right at $5.0\,m\,s^{-1}$ when it strikes a stationary ball. The first ball rebounds at $3.0\,m\,s^{-1}$.

Calculate:

a) the change in momentum of the first ball
b) the change in momentum of the second ball.

Solution

a) $m = 4.0\,kg$; $u = 5.0\,m\,s^{-1}$; $v = -3.0\,m\,s^{-1}$ (note that this velocity is negative because it is the opposite direction)

$\Delta p = mv - mu$
$\Delta p = (4 \times -3) - (4 \times 5)$
$\Delta p = -12 - 20$
$\Delta p = -32\,kg\,m\,s^{-1}$

The answer is negative because the change in momentum is from the right to the left.

b) The change in momentum of the second ball has the same magnitude but the opposite direction.

$\Delta p = +32\,kg\,m\,s^{-1}$

Force–time graphs

The force that an object experiences during a collision or explosion can be represented using a force–time graph.

The graph in Figure 3.8 is too difficult to analyse so in exam questions you may be asked about the type of graph shown in Figure 3.9.

The impulse on an object is the average force multiplied by the time of contact. The impulse can therefore be found by calculating the area under the graph. The area under the graph is therefore also equal to the change in momentum.

When a car crashes into a wall the change in momentum will be equal to the initial velocity of the car multiplied by its mass. This will have the same value no matter what happens during the collision, but the force acting on the car can be reduced if the time of the collision is increased. The larger the time, the smaller the force as $Ft = mv - mu$. This is why

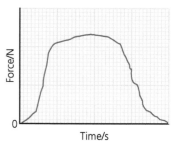

Figure 3.8 A force–time graph

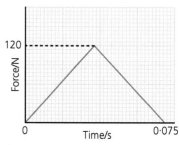

Figure 3.9 A force–time graph

cars are fitted with crumple zones. These increase the time of the collision and so reduce the size of the force on the car. Motorcycle helmets have soft inner linings for the same reason.

Key points

* The momentum of an object is its mass multiplied by its velocity.
* Momentum is a vector so it is important to give a direction when describing it.
* The unit for momentum is $kg\,m\,s^{-1}$.
* During a collision or explosion (if there are no external forces) the total momentum before is equal to the total momentum after. This is called the Law of Conservation of Linear Momentum.
* There are two types of collisions: elastic and inelastic.
* In an elastic collision, kinetic energy is conserved.
* In an inelastic collision, kinetic energy is lost (usually as heat and sound).
* In all collisions total energy is conserved.
* During an explosion, stored energy (usually in a spring or a chemical) is converted to kinetic energy.
* During an explosion or collision, the objects obey Newton's Third Law of Motion, so that the force one object exerts on the other is equal in size but opposite in direction to the force the second object exerts on the first.
* During a collision or explosion, objects exchange momentum. The momentum gained by one will be equal to the momentum, lost by the other.
* Collisions can be analysed using force–time graphs.
* The area under a force–time graph is called the impulse.
* The impulse is equal to the average force multiplied by the time of the collision.
* Impulse is a vector and its units are N s.
* The impulse on an object is equal to its change in momentum. This means that the area under a force–time graph is also equal to the change in momentum of the object.

Key words

Elastic – in collisions, an elastic collision is one in which kinetic energy is conserved

Explosion – two or more objects move apart during an explosion; the kinetic energy gained comes from energy stored in a spring or a chemical

Impulse – the average force multiplied by the time

Inelastic – in collisions, an inelastic collision is one in which kinetic energy is lost

Momentum – the mass of an object multiplied by its velocity

Newton's Third Law of Motion – for every action there is an equal but opposite reaction

Questions ❓

1   A car of mass 1200 kg is travelling at 12·0 m s⁻¹. Calculate the momentum of the car.

2   A ball of mass 2·8 kg travelling at 4·0 m s⁻¹ to the right along a frictionless surface collides with a stationary 1·0 kg ball. After the collision the 1·0 kg ball moves to the right with a velocity of 8·4 m s⁻¹. Calculate the velocity of the 2·8 kg ball after the collision.

3   A paintball gun of mass 2·2 kg fires a 0·2 kg pellet at 8·0 m s⁻¹. Calculate the initial recoil velocity of the gun.

4   A 1·2 kg ball is travelling at 2·0 m s⁻¹ to the right along a frictionless surface. Another ball, mass 2·0 kg, is travelling to the left at 1·6 m s⁻¹ along the same surface. The balls collide head-on. The 1·2 kg ball rebounds to the left at 1·0 m s⁻¹.
   a) Calculate the velocity of the 2·0 kg ball after the collision.
   b) Show, by calculation, whether the collision is elastic or inelastic.
   c) The time of contact between the balls is 0·15 s. Calculate the average force that the 2·0 kg ball exerts on the 1·2 kg ball.
   d) What is the average force exerted by the 1·2 kg ball on the 2·0 kg ball?

5   A medicine ball of mass 1·5 kg is dropped onto a hard wooden surface. The force–time graph for the ball is shown in Figure 3.10.
   a) Calculate the impulse on the ball.
   b) If the ball is travelling at 5·0 m s⁻¹ when it strikes the wooden surface, calculate the rebound velocity of the ball.
   c) The wooden surface is replaced by a soft foam surface. Describe and explain the changes this would produce in the shape of the force–time graph.

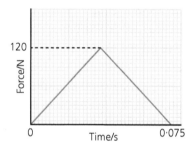

Figure 3.10 A force–time graph

6   A 3·0 kg trolley travels in a straight line towards a stationary 6·0 kg trolley as shown in Figure 3.11.
   After the collision, the trolleys move as shown in Figure 3.12. What is the speed, *v*, of the 6·0 kg trolley after the collision?
   A   0·4 m s⁻¹
   B   1·2 m s⁻¹
   C   2·4 m s⁻¹
   D   2·5 m s⁻¹
   E   3·0 m s⁻¹

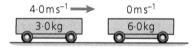

Figure 3.11 Two trolleys about to collide

Figure 3.12 Two trolleys after collision

7   Two trolleys travel towards each other in a straight line along a frictionless surface.

Figure 3.13 Two trolleys about to collide

The trolleys collide. After the collision, the trolleys move as shown below.

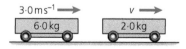

Figure 3.14 Two trolleys after collision

Which row in the table gives the total momentum and the total kinetic energy **after** the collision?

	Total momentum/kg m s⁻¹	Total kinetic energy/J
A	8	4
B	12	18
C	18	27
D	18	31
E	22	31

Table 3.1

8   The diagram shows the masses and velocities of two trolleys just before they collide on a level bench. After the collision, the trolleys move along the bench joined together.

How much kinetic energy is lost in this collision?

A   0 J

B   8·0 J

C   16 J

D   24 J

E   32 J

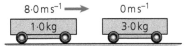

Figure 3.15 Two trolleys about to collide

29

Chapter 4

Chapter 4
Gravitation: projectiles and satellites

What you should know

★ An understanding of projectiles and satellites
★ Resolving the motion of a projectile with an initial velocity into horizontal and vertical components and their use in calculations
★ Comparison of projectiles with objects in free fall
★ Gravitational field strength of planets, natural satellites and stars
★ Calculating the force exerted on objects placed in a gravity field
★ Newton's Law of Universal Gravitation

Understanding projectiles and satellites

Projectiles

A **projectile** is an object that is released or thrown into the air. Once released, it has no external force applied to it affecting its motion other than gravity and air resistance.

Objects that are dropped or launched vertically can have their motion determined using the equations of motion. If dropped, the projectile will accelerate and behave as described earlier. If launched upwards, it will reach a maximum height then fall back to Earth in the same manner as a dropped object.

A body released from, say, 1 m above ground will fall under the influence of gravity. This force (of gravity) will cause it to accelerate. You may be required to describe how to measure the acceleration due to gravity. Here is a way in which this experiment may be described.

Example

Experiment to measure the acceleration due to gravity

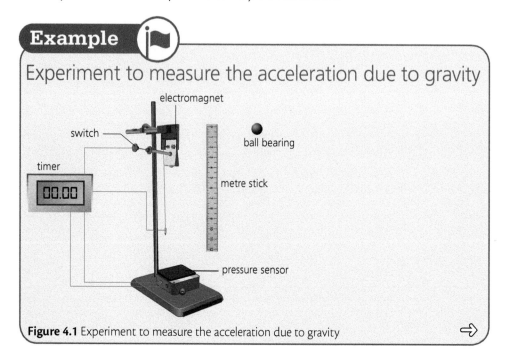

Figure 4.1 Experiment to measure the acceleration due to gravity

The electromagnet is switched on and the ball bearing is placed under the magnet.

The distance from the **bottom** of the ball bearing to the pressure sensor is measured using the metre stick.

The electromagnet is switched off. This starts the timer. The ball falls and hits the pressure sensor. This switches off the timer.

You will now have the following data:

   s – the distance between the bottom of the ball and the pressure sensor

   t – the time for the ball to fall distance, s

   u – the initial velocity of the ball ($0\,m\,s^{-1}$)

The acceleration of the ball can be calculated using $s = ut + \frac{1}{2}at^2$.

Most projectile problems involve an object that is launched either upwards or downwards, at an angle to the horizontal. The simplest one is where an object is moving horizontally and then falls off a bench, for example. The key to dealing with projectiles is to separate the horizontal and vertical motions. The horizontal motion of a projectile has no bearing on the time an object is in the air or takes to fall from a height.

Example

A ball rolls off a 1·7 m high bench as shown in Figure 4.2.

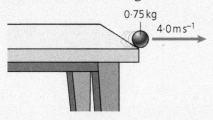
0·75 kg
4·0 m s⁻¹

Figure 4.2 A ball rolling off a bench

a) Calculate how long the ball takes to reach the ground.
b) Calculate how far away from the bench the ball lands.
c) Calculate the vertical component of its velocity as it hits the ground.
d) Calculate the angle at which the ball hits the ground.
e) Calculate the velocity at which the ball hits the ground.

Solution

a) Vertical motion: $v = ?$

   $u = 0\,m\,s^{-1}$ (no vertical velocity)

   $a = 9.8\,m\,s^{-2}$ (−ve not needed as all motion is downwards)

   $t = ?$

   $s = 1.7$

To calculate t we use $s = ut + \frac{1}{2}at^2$

$u = 0\,s$, so $s = \frac{1}{2}at^2$

rearranging gives

$$t = \sqrt{2 \times \frac{1\cdot7}{9\cdot8}}$$

$t = 0\cdot59\,s$

b) Horizontal motion:

$s = v \times t = 4\cdot0 \times 0\cdot59 = 2\cdot36 = 2\cdot4\,m$

c) Using the equations of motion:

$v = u + at = 0 + 9\cdot8 \times 0\cdot59 = 5\cdot78 = 5\cdot8\,m\,s^{-1}$

d) We combine the two vectors: $v_{hor} = 4\,m\,s^{-1}$; $v_{vert} = 5\cdot8\,m\,s^{-1}$

The angle $Z°$ is given by $\tan Z° = 5\cdot8/4\cdot0 = 1\cdot45$, so $Z° = 55\cdot4° = 55°$ to the horizontal.

e) We combine the velocities using a scale diagram or Pythagoras. This gives a resultant velocity of $\sqrt{4^2 + 5\cdot8^2} = 7\,m\,s^{-1}$.

More complicated examples involve projectiles which have an initial velocity at some angle to the horizontal. To deal with these we need to resolve the initial velocity into its horizontal and vertical components.

Example

A golf ball is struck and leaves the ground. It has an initial velocity of $25\,m\,s^{-1}$ at an angle of $14°$ to the horizontal. The initial horizontal and vertical components of this velocity can be calculated as follows:

vertical velocity, $v_v = 25 \times \sin 14° = 6\cdot0\,m\,s^{-1}$

horizontal velocity, $v_h = 25 \times \cos 14° = 24\cdot3\,m\,s^{-1} = 24\,m\,s^{-1}$

The vertical velocity is then used along with the equations of motion to calculate what is asked from the question. We do this using the method shown earlier with v, u, a, t and s.

The horizontal velocity is used to determine the distance the ball travels once we have determined the length of time it is in the air.

Using the information above, calculate the maximum height achieved by the golf ball.

Solution

When the ball is at its maximum height (zenith) its vertical component of velocity is zero. It stops moving upwards prior to falling downwards. We can use this in our equations of motion:

$v = 0\,m\,s^{-1}$

$u = (+)6\cdot0\,m\,s^{-1}$

$a = -9\cdot8\,m\,s^{-2}$

$t = ?$

$s = ?$

The question asks for the height of the ball, s. We can use $v^2 = u^2 + 2as$. Rearranging gives

$$s = \frac{v^2 - u^2}{2a} = \frac{0^2 - 6 \cdot 0^2}{2 \times (-9 \cdot 8)} = \frac{-36}{-19 \cdot 6} = 1 \cdot 8\,\text{m (a low pitch and run)}$$

Hints & tips

Remember the following information when you are working with projectiles.

✓ *When two objects are projected horizontally from the same height, they will take the same time to reach the ground irrespective of their horizontal velocity.*

✓ *The mass of a projectile has no effect on its motion.*

✓ *For a projectile projected vertically, or at an angle upwards, its vertical velocity at its maximum height is zero.*

✓ *At the instant a projectile is travelling at 45° to the vertical, the horizontal and vertical components of its velocity have the same size.*

Satellites

Objects dropped vertically and objects launched horizontally have identical vertical motions.

Satellites are projectiles that are travelling so quickly in a horizontal direction that they do not 'fall' to the ground. The Earth curves 'downwards' at the same rate the satellites do, so they remain at the same height above the surface of the Earth. Satellites stay in orbit around the Earth because they are at the correct velocity in relation to their height above the Earth. Low-orbit satellites travel more quickly than higher satellites.

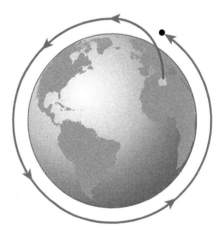
Figure 4.3 Satellite orbits around the Earth

Free fall

Free fall is a term used for the motion of a body where its weight is the only force acting on it. It is often used alongside terms like 'weightlessness' to describe certain situations. If we ignore air resistance, a falling ball, for example, could be said to be in free fall. A spacecraft orbiting the Earth is in free fall as the only force acting on it is the Earth's gravitational attraction. The Moon is in free fall as its orbit is determined by the gravitational attraction between the Earth and the Moon.

Astronauts in the International Space Station (ISS) are in free fall as they are falling at the same rate as the ISS. As a result they appear weightless. The ISS is 'falling' away from them at the same rate as they are falling. If this is confusing, consider the idea of a person in a lift, where the person is falling at the same rate as the floor of the lift. An astronaut appears weightless because the floor is not supporting them. This leads to confusion as people watching astronauts 'float' believe that this is caused by the lack of gravity in space.

Gravity and mass

The force of attraction that objects with mass experience is referred to as **gravity**. This is a naturally occurring phenomenon. In cosmological terms, gravity is responsible for the dispersed matter in the Universe being able to coalesce and bind or remain together. This leads to the formation (over very long periods of time) of stars, planets and galaxies. Gravity is also responsible for keeping the planets and other objects that make up our Solar System in orbit around the Sun.

Objects with mass experience this attractive force from other objects with mass (massive). Large objects have a correspondingly larger **gravitational field strength**, g. On Earth, g is approximately 9.8 N kg^{-1} but this varies slightly depending upon where on the Earth's surface g is measured.

This means that an object with a mass of 1 kg experiences a force of 9.8 N due to gravitational attraction. This force is what we call the object's 'weight'. A 10 kg mass has a weight of 98 N.

The weight of an object can be calculated using the relationship $W = m \times g$. We can use this relationship to calculate the weight of objects on other bodies if we can calculate or measure g for other planets and satellites.

Newton's Law of Universal Gravitation

Sir Isaac Newton derived a formula for the force of attraction between any two objects:

$$F = G\frac{m_1 m_2}{r^2}$$

where

   F is the force between the masses

   G is the **gravitational constant**; its value is $6.67 \times 10^{-11} \text{ m}^3 \text{ kg}^{-1} \text{ s}^{-2}$

   m_1 is the mass of one object

   m_2 is the mass of the second object

   r is the distance between the masses.

Objects in the Solar System experience gravitational effects from more than just one body. The Earth orbits the Sun but our orbit 'wobbles' slightly due to the gravitational effects of the Moon orbiting around us.

Example

Newton's formula can be used to calculate the gravitational field strength of a planet.

The gravitational field strength is defined as the force per unit mass. If we use the radius of the planet as the value for r, the mass of the planet as m_1 and 1 kg as m_2 then we can find the gravitational field strength.

For example, for Jupiter:

   $r = 7.15 \times 10^7 \text{ m}$

   $m_1 = 1.9 \times 10^{27} \text{ kg}$

   $m_2 = 1 \text{ kg}$

$$F = G\frac{m_1 \times m_2}{r^2} = 6{\cdot}67 \times 10^{-11}\frac{1{\cdot}9 \times 10^{27} \times 1}{(7{\cdot}15 \times 10^7)^2}$$

$$= 24{\cdot}8\,\text{N}$$

This is the force on 1 kg so the gravitational field strength on Jupiter is $24{\cdot}8\,\text{N}\,\text{kg}^{-1}$.

Example

Experiment to measure the universal gravitational constant, *G*

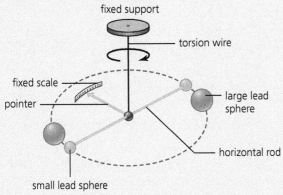

Figure 4.4 Apparatus for measuring the universal gravitational constant, *G*

Two small lead spheres are separated from each other by a rigid bar and the bar is supported by a torsion wire. It is possible to calibrate the property of the torsion wire so that we know the magnitude of the force that will rotate the wire through a certain angle.

Two large lead spheres are brought close to the small spheres. There will be a force of attraction (gravitational) between each small and large sphere. These forces on the spheres will cause the wire to twist and the angle it twists through can be measured using the fixed scale. We can then use this to determine the force between the spheres.

We can measure the mass of each of the spheres and their 'centre-to-centre' separation. This means that the only quantity in the equation we do not know is *G* and so it can be calculated.

Example

Calculations using Newton's formula give very small answers for most objects.

For example, we can calculate the force of gravitational attraction between two 50 kg students who are sitting 2·0 m apart:

$$F = G\frac{m_1 \cdot m_2}{r^2} = 6{\cdot}67 \cdot 10^{-11}\frac{50 \cdot 50}{2^2}$$

$$= 4{\cdot}2 \times 10^{-8}\,\text{N}$$

This is a very small force. The only time that this equation produces large numbers is when at least one of the objects is a planet, moon or a star.

Key points

* A projectile is an object whose motion is affected by a gravitational field.
* The motion of a projectile is analysed by splitting its motion into horizontal and vertical components.
* If air resistance is **negligible**, there is no unbalanced force in the horizontal direction. This means there is no acceleration in the horizontal direction.
* In the vertical direction, the projectile accelerates due to gravity. The equations of motion are therefore used to analyse the vertical motion of the projectile.
* Whether a projectile is moving upwards or downwards, the vertical force acting on it is downwards, so its acceleration is always downwards.
* The vertical motion of a projectile is identical to the free-fall motion of an object released from the same height with the same vertical velocity.
* A satellite is an object that is in an orbit around another, usually a planet.
* Satellites follow curved paths because they are projectiles. They have a constant horizontal speed but are accelerating in the vertical direction. The curvature of the planet matches the curvature of their path so they remain the same height above the planet.
* The gravitational field strength of a planet, natural satellite or star is the weight per unit mass at the object's surface.
* The force an object experiences in a gravitational field is called its weight. The weight is calculated using the equation $W = mg$.
* Newton's Universal Law of Gravitation allows us to calculate the force of attraction between any two massive objects. In practice, we would normally use this to calculate the force of attraction between planets, stars and natural satellites, but it can be used for any two massive objects.

Key words

Gravitational field strength – the weight per unit mass
Negligible – so small that it can usually be ignored
Projectile – an object that has both a horizontal and vertical component to its motion
Satellite – an object in orbit around another

Questions

1   An object is launched with a velocity of 22 m s⁻¹ at an angle of 15° to the horizontal.
   a) Calculate the horizontal and vertical components of the initial velocity.
   b) How long will it take for the object to reach its highest point?
   c) Calculate the maximum height reached by the object.
   d) Calculate the horizontal distance travelled by the object.

2   A ball is rolled off a horizontal bench with a velocity of 3·6 m s⁻¹. The bench is 1·2 m above the floor.
   a) Calculate the horizontal distance it travels before hitting the ground.
   b) Another ball is rolled off the bench at a velocity of 7·2 m s⁻¹. Explain why it strikes the floor at a distance of 3·6 m from the bench.

3   A ball is projected from the ground with a velocity of 12 m s⁻¹ at 60° to the horizontal. Calculate:
   a) the initial horizontal component of velocity
   b) the initial vertical component of velocity
   c) the time taken for the projectile to reach its maximum height
   d) the maximum height reached
   e) the range of the projectile.   ⇨

4 Explain why a satellite that is travelling at a constant speed is accelerating.

5 A projectile is fired with an initial horizontal velocity, u. After a time its velocity is $10\,\mathrm{m\,s^{-1}}$ at $45°$ to the vertical. Determine the initial horizontal velocity.

Use these data in the following questions:

$G = 6.67 \times 10^{-11}\,\mathrm{N\,m^2\,kg^{-2}}$

Mass of the Earth $= 6.0 \times 10^{24}\,\mathrm{kg}$

Mass of the Moon $= 7.3 \times 10^{22}\,\mathrm{kg}$

Radius of the Earth $= 6.4 \times 10^{6}\,\mathrm{m}$

Radius of the Moon $= 1.7 \times 10^{6}\,\mathrm{m}$

Radius of the Moon's orbit around the Earth $= 3.84 \times 10^{8}\,\mathrm{m}$

6 Two dust particles each have a mass of $0.1\,\mathrm{g}$. Calculate the force of gravitational attraction between them when they are:
   a) $1.0\,\mathrm{m}$ apart
   b) $1.0\,\mathrm{mm}$ apart.

7 Calculate the gravitational force of attraction between the Earth and the Moon.

8 Calculate the gravitational field strength at the surface of the Moon.

9 A $200\,\mathrm{kg}$ mass is in orbit above the Earth. In this orbit its weight is $200\,\mathrm{N}$. Calculate the height of the satellite above the surface of the Earth.

*10 A basketball player throws a ball with an initial velocity of $6.5\,\mathrm{m\,s^{-1}}$ at an angle of $50°$ to the horizontal. The ball is $2.3\,\mathrm{m}$ above the ground when released.

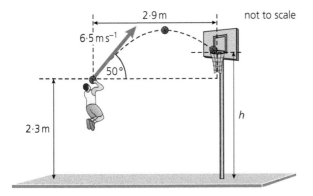

Figure 4.5 Basketball player throwing a ball

The ball travels a horizontal distance of $2.9\,\mathrm{m}$ to reach the top of the basket. The effects of air resistance can be ignored.

   a) Calculate:
      (i) the horizontal component of the initial velocity of the ball
      (ii) the vertical component of the initial velocity of the ball.
   b) Show that the time taken for the ball to reach the basket is $0.69\,\mathrm{s}$.
   c) Calculate the height, h, of the top of the basket.
   d) A student observing the player makes the following statement:
      'The player should throw the ball with a higher speed at the same angle. The ball would then land in the basket as before but it would take a shorter time to travel the 2.9 metres.'
      Explain why the student's statement is incorrect.

*11 A student investigates the motion of a ball projected from a launcher. The launcher is placed on the ground and a ball is fired vertically upwards. The vertical speed of the ball as it leaves the top of the launcher is 7·0 m s⁻¹. The effects of air resistance can be ignored.

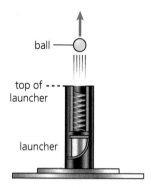

Figure 4.6 Ball projected from a launcher

a) (i) Calculate the maximum height above the top of the launcher reached by the ball.
  (ii) Show that the time taken for the ball to reach its maximum height is 0·71 s.
b) The student now fixes the launcher to a trolley. The trolley travels horizontally at a constant speed of 1·5 m s⁻¹ to the right. The launcher again fires the ball vertically upwards with a speed of 7·0 m s⁻¹.

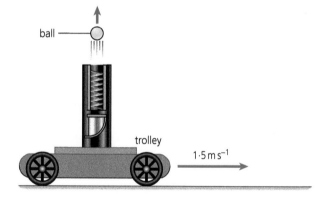

Figure 4.7 Launcher on a trolley firing a ball

(i) Determine the velocity of the ball after 0·71 s.
(ii) The student asks some friends to predict where the ball will land relative to the moving launcher. They make the following statements.

Statement X: The ball will land behind the launcher.
Statement Y: The ball will land in front of the launcher.
Statement Z: The ball will land on top of the launcher.

Which of the statements is correct? You must justify your answer.

Special relativity

Frames of reference

When we make a measurement of a velocity we do it relative to a **frame of reference**. Quite often we do not realise that we are doing this. For example, when we measure the velocity of a trolley rolling down a ramp, we are actually measuring the velocity of the trolley relative to the room. In doing so we assume the room is not moving.

If we carry out the trolley experiment on a train, we would get the same answer for the velocity of the trolley relative to the train whether the train is moving or stationary. However if we measured the velocity of the trolley relative to an observer standing beside the track, we would get a different answer depending on whether the train was moving or stationary.

This is called **Galilean invariance** and it refers to the principle that within a certain frame of reference, the relative motion of objects is the same.

During the nineteenth century, experiments carried out into the speed of light produced an unexpected result: the speed of light in a vacuum was the same for all observers regardless of the observer's motion. In other words, if you are travelling towards a beam of light, you measure the same speed for the light as you do when you are travelling away from the light. The light from a fast-moving jet, for example, left the jet at the same speed as light from a stationary object.

This result was unexpected. It seemed obvious that the relative motion between an observer and a light source would affect the measured speed of the light, but it did not.

Time dilation

Imagine you are on a moving train. There is a light source on the floor and a mirror on the roof. A pulse of light is sent from the source to the mirror and back down to the floor vertically. The time taken, t, can be measured. This is the time for an observer moving with the object (that is, the observer on the train).

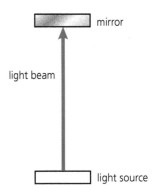

Figure 5.1 Light beam from source to a mirror on the roof

Now imagine you observe the same event from a stationary position at the side of the track. As the train is moving, you would see the light travel a further distance as it is no longer travelling vertically but diagonally. The speed of light is the same for all observers so the stationary observer must measure a longer time, t', than the moving observer.

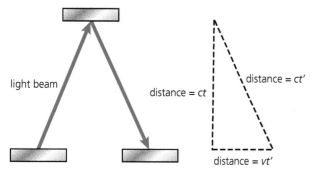

Figure 5.2 Light beams and mirrors

This means that time elapses differently for different observers depending on their relative velocity.

This idea is expressed in the following formula:

$$t' = \frac{t}{\sqrt{1 - \left(\frac{v}{c}\right)^2}}$$

where

t' is the observed time for the stationary observer

t is the observed time for the moving observer

v is the velocity of the observed object

c is the velocity of light.

It is important to realise that the difference in time recorded by a stationary and moving observer only becomes significant if the relative velocity is very large. The effects at velocities of less than 10% of the speed of light are usually negligible. We often describe velocities of greater than 10% of the speed of light as **relativistic velocities**.

Time dilation was confirmed by measurement when the number of **muons** reaching the surface of the Earth was compared to the number that would be expected from calculations that ignore the time dilation effects. Muons are unstable subatomic particles produced when cosmic rays enter the Earth's atmosphere from space at a height of 10 km. Muons have a half-life of 1·56 µs and travel towards the Earth with a velocity of 98% of the speed of light.

If we ignore any relativistic effects, the time taken for muons to travel to the surface of the Earth is measured as 34 µs. This represents nearly 22 half-lives and so very few muons should be detected at the Earth's surface.

When we calculate the time from the moving muon's frame of reference, we find that to reach the ground they have been travelling for 6·8 µs.

This represents just over four half-lives. This would mean that many more muons should be detected at the surface of the Earth than would be expected if relativistic effects are ignored.

When the actual number of muons reaching the surface of the Earth is measured, it is found to agree with the number predicted by the time-dilation formula.

Example

A spaceship travels at a velocity of $0.750c$ on an exploratory mission. It returns to Earth 22 years later (as measured by observers on Earth). How many years have passed for the astronauts?

Solution

$$t' = \frac{t}{\sqrt{1 - \left(\frac{v}{c}\right)^2}}$$

$$t' = \frac{t}{\sqrt{1 - 0.75^2}}$$

$$t' = \frac{t}{\sqrt{0.5625}}$$

$$t' = \frac{t}{0.661}$$

$22 \times 0.661 = 14.6$ years

Length contraction

Objects that are travelling at relativistic velocities appear to be shorter to a stationary observer than they do to an observer moving at the same velocity as the object. Time does not elapse at the same rate for the stationary observer as for the moving observer. When they use their respective times to calculate the length of the object, the stationary observer obtains a shorter length for the object than the moving observer.

The **length contraction** is given by the formula:

$$l' = l \times \sqrt{1 - \left(\frac{v}{c}\right)^2}$$

where

    l' is the length measured by the stationary observer

    l is the length measured by the moving observer

    v is the velocity of the observed object

    c is the velocity of light.

Hints & tips

The length that the stationary observer measures is always less than the moving observer measures. As with the calculations for time dilation, if you get a longer length for the stationary observer, you must have done something wrong. You need to review what you have done and check your calculations.

Example

A meteorite of length 125 m is travelling through space at 0.25 c. Calculate its apparent length when measured by a stationary observer.

Solution

$$l' = l \times \sqrt{1 - \left(\frac{v}{c}\right)^2}$$

$$l' = 125 \times \sqrt{1 - 0.25^2}$$

$$l' = 125 \times \sqrt{1 - 0.0625}$$

$$l' = 125 \times \sqrt{0.9375}$$

$$l' = 121\,m$$

Key points

* The speed of light is the same for all observers in spite of their motion.
* The laws of motion are true in all frames of reference. This is called Galilean invariance.
* Because the speed of light is constant for all observers, two observers viewing the same event will measure different times and distances depending on how they are moving.
* If an observer in a stationary frame of reference and an observer in a moving frame of reference measure the time for the same event, the observer in the stationary frame of reference records a larger time. This is called time dilation.
* The length of a moving object measured by a stationary observer will be less than that measured by an observer moving with the object. This is called length contraction.
* The time between events measured by a stationary observer will be greater than that measured by an observer moving with the events. This is called time dilation.
* The theoretical predictions made by the theory of special relativity can be checked by experiment, for example, by measuring the number of muons reaching the surface of the Earth.

Key words

Frame of reference – the background against which measurements are made

Invariance – the laws of physics are the same for all observers in all frames of reference

Muon – a subatomic particle with a short half-life

Questions ?

1   A spacecraft is travelling at $0.20c$ relative to the Earth. A clock on the spacecraft measures the time for the spacecraft to travel between two points as 10.0 s. Calculate the time elapsed for an observer on the Earth.

2   A space probe travels from the Earth towards the Sun and back at 10.0% of the speed of light. The total distance travelled is 2.10×10^{11} m. The probe length is measured as 20.0 m on board the probe during its flight. Calculate:

   a) the time elapsed for a stationary observer on the Earth viewing the probe's journey

   b) the time elapsed for a clock placed on the space probe during its journey

   c) the length of the probe observed by a stationary observer on the Earth.

Chapter 6
The expanding Universe

The Doppler Effect

The **Doppler Effect** is the name given to a naturally occurring phenomenon which alters the relative **frequency** of radiation that we can detect. It is noticeable in sound waves where we hear a slightly higher frequency when a source is travelling towards us and a lower frequency when that source is travelling away from us.

With sound, we hear a slightly higher pitch from a source heading towards us as each successive wave is generated from a slightly closer distance. This causes the waves to arrive more frequently, increasing the pitch. The opposite happens for a sound source travelling away from us.

The observed frequency for a stationary observer is given by:

$$f_o = f_s \left(\frac{v}{v \pm v_s} \right)$$

where

f_o = observed frequency
f_s = source frequency
v = wave velocity
v_s = source velocity.

We can only hear the difference in frequency when the moving object is moving relatively quickly compared to the velocity of sound. Listening to people walking to or from us does not alter the frequency of their voice enough for us to notice. With ambulances and aeroplanes, however, there is a discernible difference.

Hints & tips ⭐

Care needs to be taken when using this equation. When the source is moving towards the observer, you should subtract the velocity of the source (this will give a bigger frequency). When the source is moving away, you should add the velocity of the source (giving a smaller frequency).

Example

A person in a car hears the siren of an approaching ambulance. The frequency of the siren is 1200 Hz. The ambulance is heading towards the person at 22 m s^{-1}.

Calculate the frequency heard by the person in the car.

Solution

$$f_0 = f_1 \left(\frac{v}{v - v_s} \right)$$

$$= 1200 \times \left(\frac{340}{340 - 22} \right)$$

$$= 1200 \times \frac{340}{318}$$

$$= 1283 \, \text{Hz}$$

The red shift

This effect can also be observed in moving objects that emit light (if they are moving quickly enough). Objects that are moving quickly towards us would have their 'light' moved towards the higher-frequency end of the spectra (blue end) and objects moving away from us would appear to have their light moved towards the lower-frequency end of the spectra (red end). How much the spectra had moved towards the red end is called the **red shift**. The red shift can be used to give us an indication of how quickly a star is moving away from us.

The magnitude of the red shift is given by:

$$z = \frac{\lambda_o - \lambda_r}{\lambda_r}$$

where λ_o is the wavelength observed and λ_r is the wavelength emitted, and

$$z = \frac{v}{c}$$

where v is the relative velocity of the star and c is the speed of light.

Example

When elements are excited and their spectra analysed, we note that particular lines (at certain wavelengths) are visible. One observed line of hydrogen has a wavelength of 656 nm.

An observed galaxy was found to have the same lines but the wavelengths have shifted. The measured wavelength of the corresponding line is 675 nm

1 Calculate the red shift of the galaxy.

Solution

$$z = \frac{\lambda_0 - \lambda_1}{\lambda_0} = \frac{675 - 656}{656} = 0.029$$

2 Calculate the velocity at which the galaxy is moving away from us.

Solution

$$z = \frac{v}{c} = \frac{v}{3.00 \times 10^8}$$

$$0.029 = \frac{v}{3.00 \times 10^8}$$

$$v = 0.029 \times 3.00 \times 10^8$$

$$v = 8.7 \times 10^6 \, \text{m s}^{-1}$$

Hubble's Law

It was found by examining the spectra of a certain class of star (Cephid variables) that almost all of the galaxies in the Universe are moving away from us. Edwin Hubble also discovered that the objects further away from us are moving away more quickly. He investigated the relationship between the distance (d) to far off galaxies and their relative velocity (v) and determined a 'simple' relationship:

$$\frac{v}{d} = \text{a constant, } H_0$$

where d is the distance to the **galaxy** and v is the **recessional velocity** of the galaxy.

This is known as **Hubble's Law** and H_0 is called **Hubble's constant**. The value for Hubble's constant is being refined continually as improvements in technology allow more accurate measurements to be taken. As of 2018, the value of $H_0 = 2.30 \times 10^{-18} \, \text{s}^{-1}$.

Using Hubble's Law, calculate the distance to the galaxy in the previous example.

$$\frac{v}{d} = H_0$$

$$\frac{8.7 \times 10^6}{d} = 2.3 \times 10^{-18}$$

$$d = \frac{8.7 \times 10^6}{2.3 \times 10^{-18}}$$

$$d = 3.78 \times 10^{24} \, \text{m (about 400 million light years)}$$

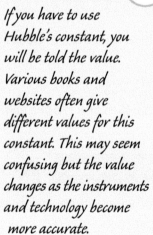

Hints & tips

If you have to use Hubble's constant, you will be told the value. Various books and websites often give different values for this constant. This may seem confusing but the value changes as the instruments and technology become more accurate.

The implications of this law are that if all galaxies are moving away from us, it would appear that the Universe is expanding. Hubble and others obtained values for the distances and relative velocities of stellar objects by measuring their brightness and red shift. If we use the equation $\frac{v}{d} = H_0$, we can calculate how long the galaxy has been moving. If we know its distance and its velocity, and since $\frac{d}{v} = \text{time}$ and $\frac{d}{v} = \frac{1}{H_0}$, it follows that $\frac{1}{H_0} = \text{time}$. This gives a value for the age of the Universe which equates to about 13 700 000 000 years.

Key points !

* The Doppler Effect allows us to measure the velocity of stars and galaxies that are distant from the Earth.
* The further away an object is from the Earth, the faster it is travelling away from us. This is Hubble's Law.
* Hubble's Law gives rise to an equation which says that the recessional velocity of the object divided by the object's distance from us is equal to a constant. This constant is known as Hubble's constant, H_o.
* Hubble's constant allows us to estimate the age of the Universe.

Key words

Doppler Effect – a naturally occurring phenomenon which alters the relative frequency of radiation that we can detect

Frequency – the number of times something happens per second; for example, the number of complete AC cycles per second or the number of waves per second

Galaxy – a large collection of stars

Hubble's Law – this law states that the further away from the Earth an object is, the faster it is moving away from the Earth; this law only applies to very distant objects

Hubble's constant – Hubble's constant is the recessional velocity of an object divided by the distance from the Earth

Recessional velocity – the velocity at which a distant object is moving away from the Earth

Red shift – the spectra of all distant objects are shifted towards the red end of the spectrum; the faster an object is travelling away from the observer, the more its spectrum is redshifted

Questions

In the following questions, take the speed of sound to be $340 \, \text{m s}^{-1}$.

1 A police car is travelling at $40 \, \text{m s}^{-1}$ towards a stationary observer. The police car's siren has a frequency of $800 \, \text{Hz}$.
   a) Calculate the frequency heard by the observer.
   b) After the police car passes the observer, it increases its speed to $50 \, \text{m s}^{-1}$. Calculate the frequency now heard by the observer.

2 A distant galaxy is $4.44 \times 10^{23} \, \text{m}$ from the Earth. Its red shift is measured to be 0.0034.
   a) Show that the recessional velocity is $1.02 \times 10^6 \, \text{m s}^{-1}$.
   b) Calculate the value that these data give for Hubble's constant in s^{-1}.
   c) Use these data to give the age of the Universe in billions of years.

Chapter 7
The Big Bang Theory

Dark matter and dark energy

Hubble's Law implies that the Universe is expanding from some event in time and space that occurred approximately 13·7 billion years ago. Before this idea became the accepted view of the origin of the Universe, a famous British astronomer, Fred Hoyle, described this single event as some sort of 'big bang' and the term has remained.

As a result of this event, or cosmological expansion, all the matter in the Universe was scattered in all directions. If we apply the rules of classical physics to this event, the matter should continue to travel at the same rate unless some other force is applied. Gravitational attraction between different bodies of matter will exert unbalanced forces throughout the Universe. It is this gravitational attraction that causes the matter to combine and form celestial objects (stars, galaxies and so on). As these objects are moving and then attracted to other objects, galaxies are pulled at various directions to their initial movement and this causes them to spin or rotate. The rate of rotation depends upon the magnitude of the masses involved and by measuring the rates we can obtain an indication of the mass of the objects. The greater the speed of rotation, the greater the mass.

Astronomers measured the rates of rotation of galaxies and calculated a figure for the mass required for this rate of rotation. This figure was greater than the amount of matter that could be observed by telescopes and other detectors. This led to the conclusion that there must exist in the Universe a type of matter which cannot be detected using instruments based on electromagnetic-spectrum detectors. This matter has been given the name **dark matter**.

Recent astronomical measurements from space-based and high-altitude telescopes have allowed astronomers to make detailed observations of very distant objects. Some of these objects were found to be not only moving away from us but *accelerating* away from us. This would suggest that the Universe's expansion is increasing! This is difficult to account for.

This observation led to the proposal that there exists a form of energy that is not detectable with current astronomical devices, similar to dark matter, and that it is this so-called **dark energy** which is responsible for the increasing rate of expansion of the Universe.

Current estimates indicate that less than 5% of the Universe's energy and matter is the material we are familiar with, such as atoms, molecules and electromagnetic radiation; approximately 70% is dark energy and 25% is dark matter.

Hints & tips

This is an area of ever-increasing knowledge in physics. As a result, the information in this section is subject to review.

The Big Bang Theory

We can observe stars and their motion from Earth by means of some very sophisticated telescopes which can detect the energy emitted by stars across the range of the electromagnetic spectrum (X-rays, gamma rays, microwaves and so on).

Radiation from a star (across a range of wavelengths) can be measured and from this information we can produce a graph of the intensity of radiation against wavelength of radiation. Figure 7.1 shows an example of the sort of graph that can be produced, but there are many more.

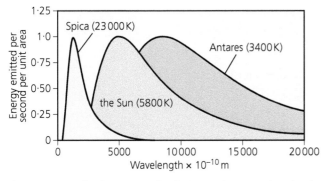

Figure 7.1 Graph of intensity of radiation against wavelength of radiation for stars

There was found to be a peak intensity at a certain wavelength followed by a gradual fall in intensity as the wavelength increased. It was discovered that the wavelength of the peak intensity is a direct indicator of the temperature of the star and that hotter stars have a smaller peak intensity wavelength. This is related to the equation $E = hf$. The energy of the radiation is proportional to its frequency and high energy indicates higher frequency, which results in a smaller wavelength.

Hints & tips

Remember that for electromagnetic waves, as the wavelength decreases, the frequency increases.

This allows us to analyse the radiation emitted by a star and from that determine the surface temperature of that star. Stars with a smaller **peak wavelength** are hotter than stars with a greater peak wavelength. Very hot stars have a blue colour, average stars like our Sun are yellow and cool stars are **red dwarfs**.

Our Sun is predicted to continue in its current state for a few more years (about 7 billion) and then as it runs out of fuel it will cool, become red and expand to form a **red giant**.

Evidence for the Big Bang Theory

Hubble's observations led scientists to consider that there was a single event from which the Universe came into being. Physicists tried to confirm this by investigating whether there were any other observable phenomena which would be indicators of this event. The sections below summarise these phenomena.

Cosmological red shift

Hubble's observations that all galaxies are moving away from us and that their distances and the speeds at which they are moving are all related is a strong indicator.

Cosmic microwave background radiation (CMBR)

If we consider that all the energy and matter originated from a central event, then the temperature at the beginning would be incredibly high. This hot, dense, gaseous beginning would emit energy because of the vast temperature (not unlike that of the stars we referred to earlier). As the expansion continued, the Universe would have changed to allow energy to propagate through space. This energy should now be in the microwave region of the electromagnetic spectrum and should appear to be the same throughout the Universe (in other words, not clumped or in regions where it is more or less concentrated). Additionally, the 'temperature' of the microwave energy should be about −270 °C or 3 K. Recent, detailed measurements of the microwave distribution throughout the Universe are in very close agreement with the theory and are taken by most scientists as proof.

The amount of hydrogen and helium in the Universe

When all matter was formed shortly after the Big Bang, the temperature was so great that it would allow only the simplest particles to form. The simplest atoms are hydrogen and helium. As this matter spread throughout the Universe, there would be areas where there was enough matter for stars to form. In these stars, heavier elements can be formed but, given the vast size of space, you would expect hydrogen and helium to be the most abundant. Scientists predicted that about 80% and 20% of all the matter in the Universe would be hydrogen and helium, respectively. Current measurements put the figures at 75% and 24%.

These pieces of evidence, in particular CMBR, all point to the Big Bang Theory being correct. There are a number of other observations which support the theory, however, these are three of the strongest.

Key points

* One of the conclusions reached from Hubble's Law is that the Universe is expanding.
* The mass of a galaxy can be estimated by measuring the orbital speed of the stars around the galactic centre.
* The total mass of the visible objects in the Universe is less than the estimated total mass of the Universe. This has given rise to the suggestion there is a large amount of matter that we cannot observe. This is called dark matter.
* The rate of expansion of the Universe is increasing. There is not enough visible energy in the Universe to account for this, so a type of energy known as dark energy has been proposed to account for this.
* Stars emit light over a range of wavelengths. The wavelength that has the highest irradiance is called the peak wavelength.
* When the spectrum of light from stars is examined it is found that the hotter the star, the shorter the peak wavelength of light that is emitted.
* The hotter the star, the higher the radiation per square metre that is emitted from the star.
* The cosmic microwave background radiation is evidence that supports the Big Bang Theory of the origin of the Universe.

Key words

Cosmic microwave background radiation – the observed background radiation that is present in every direction in the Universe; it provides strong evidence to support the Big Bang Theory of the Universe

Dark energy – a type of energy proposed to account for the missing energy in the Universe; this is to account for the fact that the rate of expansion of the Universe is increasing

Dark matter – a type of matter proposed to account for the missing mass of the Universe; this mass cannot be observed by conventional telescopes

Peak wavelength – the wavelength in a star's spectrum that has the highest irradiance

Questions

1 Which type of star has a higher temperature: red giants or blue dwarfs?
2 What happens to the radiation per unit surface area of a star as its temperature increases?
3 What has happened to the temperature of the Universe as it has expanded?
4 List three pieces of evidence that support the Big Bang Theory for the formation of the Universe.

Section 2 Particles and waves

Chapter 8
Electric fields

What you should know

★ Fields exist around charged particles and between charged parallel plates
★ Examples of electric field patterns for single-point charges, systems of two-point charges and between parallel plates
★ Movement of charged particles in an electric field
★ The relationship between potential difference, work and charge gives the definition of the volt
★ Calculation of the speed of a charged particle accelerated by an electric field
★ A moving charge produces a magnetic field
★ The determination of the direction of the force on a charged particle moving in a magnetic field for negative and positive charges (right-hand rule for negative charges)
★ The basic operation of particle accelerators in terms of acceleration, deflection and collision of charged particles

Charge and the electric field

Charge is a physical property of matter and is a characteristic property of many subatomic particles.

Charged particles interact with other charged particles and they can apply a force to each other: they can attract or repel. The magnitude of the force they apply is dependent upon the magnitude of the charge and the distance between the charged objects. We explain the interaction by using the concept of a field.

An **electric field** is the region of space around a charge and we can describe this field by the use of lines around the charge. These lines indicate what force a positive charge would experience if it was to be placed at any point.

A positive charge placed near to the charge in Figure 8.1 will experience a force acting on it to move it outwards (repulsion).

Figure 8.1 Electric field lines around a single positive point charge

A positive charge placed near to the charge in Figure 8.2 will experience a force acting on it to move it towards the other charge (attraction).

The field lines indicate the direction of the force on a positive charge but also give an indication of the magnitude. The closer together the lines, the greater the force. The **electric field strength** in this area is stronger. This can be seen if a charge moves closer to another charge; the lines are closer together, indicating that the electric field strength is greater, leading to a larger force (in other words, there is greater attraction or repulsion).

Electric fields exist around any object that is charged, not just individual charges. We can apply a charge to any object. In physics it is useful to have metal plates charged. The field between parallel metal plates is shown in Figure 8.3. This arrangement gives us an effective way of being able to move charged objects between these plates.

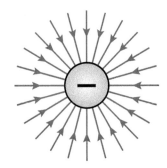

Figure 8.2 Electric field lines around a single negative point charge

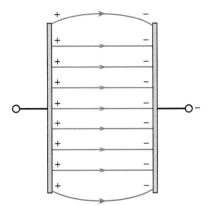

Figure 8.3 Electric field lines between oppositely charged parallel plates

As indicated earlier, a charge experiences a force when interacting in an electric field. This force varies directly with the magnitude of the charge and the strength of the electric field.

An unbalanced force will cause an object to accelerate. Charged particles tend to have a very small mass and, as a result, the acceleration of these objects is very great ($a = \dfrac{F}{m}$). The result of this is that these small particles can reach high velocities in very short distances. An electron, for example, can reach a velocity of $10^7 \, \mathrm{m\,s^{-1}}$ over a distance of a few centimetres.

A negatively charged particle placed between two charged plates will experience a force to one of the plates and will accelerate towards it as shown in Figure 8.4.

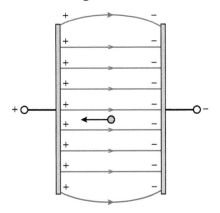

Figure 8.4 The particle is accelerating towards one of the charged plates

If the charged particle enters from the top end of the plates and is travelling vertically, it will experience a force attracting it to one side. Depending upon the initial velocity and charge of the particle, it will be deflected to one plate or the other. This allows us to direct the particle to where we want it to go. This is very useful in a number of scientific applications.

Definition of the volt

Moving charges within electric fields involves a number of physical concepts. We know that a charge will experience a force in an electric field and that this will result in movement. The energy required for this movement must come from an external source such as an electrical supply. This leads us to the idea of **electrical potential**. This is defined as the amount of energy required to move a charge of one coulomb to a point within the field.

The energy required to move a charge from a point (for example, A) to a second point (B) depends on the relative potential of points A and B. This leads us to the term **potential difference**, which refers to the difference in potential between the two points. Potential difference is measured in **volts**:

$$V = \frac{E_w}{Q}$$

where

V is the difference in potential between two points

E_w is the energy supplied to the charge

Q is the magnitude of the charge.

The term potential difference is used when discussing electric circuits where charges move and energy is transferred to various components.

Speed of charged particles

When a charge is placed between two charged parallel plates, for example, it will experience a force. This force causes the charge to accelerate from one plate to another. In order to do this, energy is required. The electrical energy from the supply is used to make the charge accelerate and increase its speed.

We use the principle of conservation of energy to calculate the velocity of the charge. The energy gained by the charge is given by $E_w = QV$. This is transformed to kinetic energy, which is calculated using $E_k = \frac{1}{2}mv^2$.

Example

A pair of plates has a potential difference of 4500 V. A proton is placed on the positive plate.

The charge of a proton is 1.6×10^{-19} and the mass of a proton is 1.67×10^{-27}.

Calculate:

a) the energy gained by the proton

b) the velocity of the proton as it reaches the opposite plate.

Solution

a) Energy gained, $E_W = QV$
$$= 1.6 \times 10^{-19} \times 4500$$
$$= 7.2 \times 10^{-16} \text{J}$$

b) $E_k = \frac{1}{2}mv^2$
$$7.2 \times 10^{-16} = \frac{1}{2} \times 1.67 \times 10^{-27} \times v^2$$
$$\text{Rearranging for } v^2 = \frac{7.2 \times 10^{-16}}{\frac{1}{2} \times 1.67 \times 10^{-27}}$$
$$v^2 = 8.62 \times 10^{11}$$
$$v = 9.3 \times 10^5 \text{m s}^{-1}$$

Magnetic fields

You should be familiar with magnets and their effects. The area of space around a magnet can be described using the concept of a **magnetic field**. This is a region in which **ferrous** materials can be influenced. Diagrams showing magnetic field lines are very similar to those of electric fields.

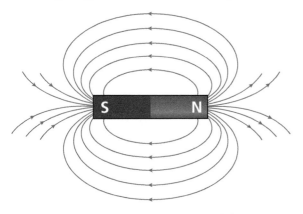

Figure 8.5 The magnetic field around a single bar magnet

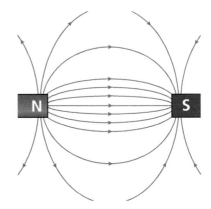

Figure 8.6 The magnetic field between two opposite poles

A stationary charge has an electric field around it. When a charge moves, however, a magnetic field is created and this is an important effect.

When a current is created in a wire, the moving charges (electrons) cause a magnetic field around the wire and it is important to be able to describe this. A current-carrying wire has a magnetic field around it which can be described using a series of concentric circles. In explaining these, there are a number of rules but these are mainly defined in terms of conventional current flow.

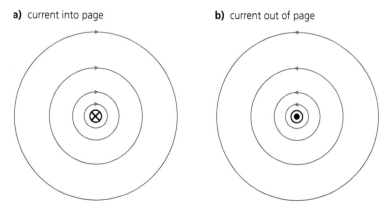

a) current into page   **b)** current out of page

Figure 8.7 The 'x' shows that the current is going into the page, while the dot represents current coming out of the page.

When drawing diagrams such as those shown in Figure 8.7, it is important to determine the direction of the field lines. This can be done using the **right-hand grip rule**.

- If your thumb represents the direction of the current, the direction your fingers curl or grip will show the direction of the magnetic field lines.
- When the current is heading towards you, the field lines are anticlockwise.
- When the current is heading away from you, the field lines are clockwise.

The interaction of these fields with an existing magnetic field (from a permanent magnet, say) can result in a force acting on the wire, causing it to move. This is the principle behind the operation of an electric motor.

When a current-carrying conductor is placed in a magnetic field, it will move in the direction shown in Figure 8.9

a)   **b)**

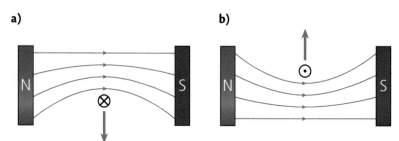

Figure 8.9 a) and b) The magnetic field of the magnets interacts with the magnetic field of the current-carrying wire.

In Figure 8.9 a) the wire is moved downwards and in Figure 8.9 b) the wire is moved upwards.

Hints & tips ⭐

You need to be careful when reading any book or set of notes about this topic. This book is written using **conventional current**, in which electricity flows from the positive terminal of a source to the negative terminal. It may be that you have been taught **electron flow**, where current flows from the negative terminal of the source to the positive. Both systems work equally well, but if you have been taught electron flow then a right-hand rule in this book becomes a left-hand rule for electron flow.

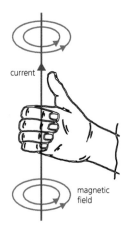

Figure 8.8 Grip the wire in your right hand with your thumb in the direction of the current. Then your fingers will curl in the direction of the magnetic field.

The direction of the force on a current-carrying wire can be determined by using the **left-hand rule**.

If we align the left hand with the middle finger in the direction of the current and the first finger being the field, then the thumb indicates the direction of force or motion. These three factors all act at right angles to each other.

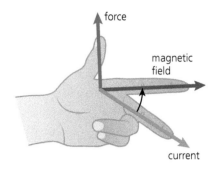

Figure 8.10 The left-hand rule

Particle accelerators

Perhaps the largest piece of scientific apparatus the world has ever built is a **particle accelerator**. The **Large Hadron Collider** is 27 km of tunnels on the France–Switzerland border with massive detectors placed at various points along its route.

A particle accelerator is a device that uses electric and magnetic fields to accelerate charged particles to very high speeds for particular purposes. Examples of particle accelerators would include cathode ray tubes that were used in older-style televisions and computer monitors.

You may have seen cathode ray tubes during your studies. They are large glass tubes that contain a particle accelerator. There are various types, for example, deflection tubes, Maltese cross tubes or Perrin tubes. A diagram of a deflection tube is shown in Figure 8.11.

All cathode ray tubes contain a particle accelerator that works in the same way. An electric field between the cathode and the anode causes electrons to be accelerated from one to the other. We can calculate the maximum kinetic energy gained by the electron due to the electric field. The charge on one electron is 1.6×10^{-19} C.

$$E = QV$$

$$E = 1.6 \times 10^{-19} \times 2.0 \times 10^3$$

$$E = 3.2 \times 10^{-16} \text{ J}$$

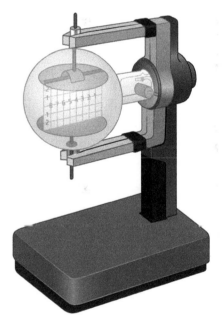

This may appear to be a very small amount of energy, but we can use it to calculate the maximum velocity of the electron as it reaches the anode. The mass of one electron is 9.11×10^{-31} kg.

$$E_k = \frac{1}{2} mv^2$$

$$3.2 \times 10^{-16} = \frac{1}{2} \times 9.11 \times 10^{-31} \times v^2$$

$$v = 2.66 \times 10^{-7} \text{ m s}^{-1}$$

Figure 8.11 A deflection tube

Despite the small amount of energy, the electrons (due to their small mass) are travelling very quickly. We can see that this type of particle accelerator is capable of producing electrons with very high velocities.

Linear accelerators (linacs)

A **linear accelerator** is a type of particle accelerator in which a charged particle is attracted towards a plate in a 'drift' tube. The particle passes through one of these tubes and is then accelerated towards the next, passes through it and is accelerated towards the next, and so on. The field between each drift tube must change rapidly so that each (new) tube attracts the particle leaving the previous tube. The particles are kept in the centre by a series of magnets but, in order to increase the energy and velocity of the particle, the accelerator needs to be long. The longest is the Stanford Linear Accelerator (SLAC) in California, USA, which is about 3 km in length. This length, along with the associated shielding and building requirements, can make the cost of a linear accelerator quite prohibitive.

Circular accelerators

A **circular accelerator**, or **cyclotron**, works by using a series of magnets to keep the particles in a circular orbit. This allows the high-frequency **voltage** supply to accelerate the particles. The circular nature of a cyclotron meant that, relative to a linear accelerator, it could be smaller in size and thus cheaper to make.

A cyclotron comprises two D-shaped sections (or 'dees') with a small gap and a very high potential difference between them. A charged particle is accelerated across the gap, then bent inside one of the dees until it is in the opposite direction. It accelerates across the gap again and the process is repeated until the particle is at the correct energy. At this point it is released for the experiment.

As the particle becomes faster, its path radius increases and it moves out towards the edge of the dees. At very high speeds, relativistic effects interfere with the efficient operation of the cyclotron and it can become difficult to adapt the fields in line with the particle.

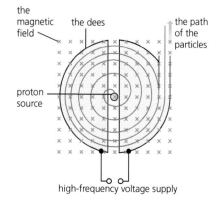

Figure 8.12 A cyclotron

Synchrotrons

There are other effects when particles are accelerated rapidly and have to travel in curved paths. Synchrotron radiation is emitted under the conditions mentioned before and this can lead to energy loss in particle physics experiments. We can build the synchrotron such that the synchrotron radiation is produced at certain frequencies for scientific and medical purposes.

The **synchrotron** is a specific type of circular accelerator where the magnetic and electromagnetic fields have been adapted to produce a very energetic and narrow ring of charged particles at very high energies. Many synchrotrons are built in order to produce radiation for experiments and medical applications rather than to accelerate particles for investigation. They can produce highly energetic photons across a range of frequencies.

Many synchrotrons produce high-energy X-rays for spectroscopy where the structure of atoms can be investigated.

The Large Hadron Collider and the Fermilab Tevatron in the USA are two examples of massive and frighteningly expensive synchrotrons which have been developed to study the structure of subatomic particles. In these facilities, particles are accelerated to incredibly high energies and then focused together in order to collide. Huge detectors are then used to detect and analyse the remnants of two protons, for example, colliding at nearly the speed of light. It is hoped that these experiments will allow us to understand the fundamental nature of particles and their interactions with each other. As detailed in Chapter 9, the Standard Model is our way of understanding or explaining how electromagnetic, weak and strong nuclear forces interact with subatomic particles. This model is undergoing continual refinement as our modern accelerators reach greater energies, which gives a greater insight into subnuclear particles.

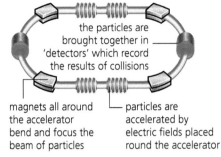

the particles are brought together in 'detectors' which record the results of collisions

magnets all around the accelerator bend and focus the beam of particles

particles are accelerated by electric fields placed round the accelerator

Figure 8.13 A synchrotron

Key points

* You should know that an electric field exists around any charged object and be able to draw the electric field pattern around a charged object.
* The arrows on the electric field lines point in the direction that a positively charged particle will move in the field.
* A charged particle in an electric field experiences a force. This force can make the particle accelerate.
* One volt is one joule per coulomb.
* The work done by an electric field on a charge is equal to the charge times the potential difference: $EW = QV$.
* You should be able to use the equations $EW = QV$ and $E_k = \frac{1}{2}mv^2$ to calculate the speed of a charged particle accelerated by an electric field.
* Moving charges produce a magnetic field.
* You should be able to predict the direction of the force on a charged particle moving through a magnetic field.
* You should be able to describe how particle accelerators work.

Key words

Electric field – the region of space around a charge
Electric field pattern – a pattern of lines around a charged object; the lines indicate the direction a positively charged particle will move in the field
Magnetic field – the area of space around a magnet
Particle accelerator – a large machine used to accelerate particles to high speeds before colliding them together
Potential difference – the number of joules per coulomb; often called voltage between two points
Voltage – another name for potential difference

Questions ?

Use these data in the following questions:

The charge on an electron = 1.6×10^{-19} C

The mass of an electron = 9.11×10^{-31} kg

The mass of a proton = 1.67×10^{-27} kg

1 An electron is placed on a negatively charged plate and is accelerated towards a positively charged plate. The potential difference between the two plates is 5000 V.
   a) Calculate the energy gained by the electron.
   b) Assuming the electron was at rest initially, calculate its speed at the second plate. You may ignore relativistic effects.
   c) A proton is now placed on the positive plate. Calculate the speed of the proton as it reaches the negative plate.

2 There is a potential difference of 2.0 kV between two points, X and Y. A proton is initially at rest at point X. The proton accelerates to point Y.
   a) Calculate:
      (i) the kinetic energy of the proton at point Y
      (ii) the maximum speed of the proton at point Y.
   b) The polarity of the supply is reversed and an electron is accelerated from rest by the 2.0 kV supply. Explain how:
      (i) the kinetic energy of the electron compares to that of the proton
      (ii) the velocity of the electron compares to that of the proton.

3 Give an advantage and a disadvantage of synchrotrons over cyclotrons.

4 The product, X, of a nuclear reaction passes through an electric field as shown.

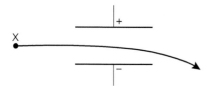

Figure 8.14 Passing through an electric field

Product X is:
A an alpha particle
B a beta particle
C gamma radiation
D a muon
E a photon.

5 An ion propulsion engine can be used to propel a spacecraft to areas of deep space. A simplified diagram of a Xenon ion engine is shown.

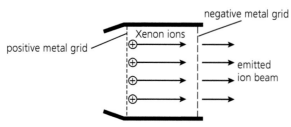

Figure 8.15 A Xenon ion engine

The Xenon ions are accelerated as they pass through an electric field between the charged metal grids. The emitted ion beam causes a force on the spacecraft in the opposite direction. The spacecraft has a total mass of 875 kg. The mass of a Xenon ion is $2 \cdot 18 \times 10^{-25}$ kg and its charge is $1 \cdot 60 \times 10^{-19}$ C. The potential difference between the charged metal grids is 2.25 kV.

a) Show that the work done on a Xenon ion as it moves through the electric field is $3 \cdot 6 \times 10^{-16}$ J.

b) Assuming the ions are accelerated from rest, calculate the speed of a Xenon ion as it leaves the engine.

*6 A student writes the following statements about electric fields.

I There is a force on a charge in an electric field.

II When an electric field is applied to a conductor, the free electric charges in the conductor move.

III Work is done when a charge is moved in an electric field.

Which of the statements is/are correct?

A I only

B II only

C I and II only

D I and III only

E I, II and III

Chapter 9
The Standard Model

What you should know

- ★ Orders of magnitude – the range of orders of magnitude of length from the very small (subnuclear) to the very large (distance to furthest known celestial objects)
- ★ The Standard Model of fundamental particles and interactions
- ★ Evidence for subnuclear particles and the existence of antimatter
- ★ Fermions, the matter particles, consist of quarks (six types) and leptons (electron, muon and tau, together with their neutrinos)
- ★ Hadrons are composite particles made of quarks: baryons are made of three quarks and mesons are made of two quarks
- ★ The force-mediating particles are bosons (photons, W and Z bosons, and gluons)
- ★ Description of beta decay as the first evidence for the neutrino

Orders of magnitude and scientific notation

When we write numbers in physics, we often use scientific notation. For example, if we want to write the distance to the Moon we can write it as 400 000 km or as 4×10^8 m. Both numbers represent the same distance, but we can say from the second form that the **order of magnitude** of the distance is 8, the same as the power when the number is written in scientific notation. The rule also applies to negative powers: the diameter of an atom is of the order of 10^{-10} m, so the order of magnitude is −10.

To give a couple of further examples, the distance to the furthest known objects in the Universe is 10^{26} m, while the diameter of an electron is of the order of 10^{-18} m.

Hints & tips

You are not expected to know the order of magnitude of every possible object. You are much more likely to be asked to compare objects or put objects into order. It may be useful to know a few distances to help you put other distances into order.

For example, the diameter of an electron is 10^{-18} m, red light has a wavelength of about 7×10^{-7} m, the length of a school lab is usually about 10 m and the distance to the Moon is approximately 4×10^8 m.

The Standard Model of fundamental particles and interactions

The story of the search for the fundamental particles and forces of nature is fascinating in its own right, but there is little room in a text such as this to do it the justice it deserves. Instead we will look at the conclusions that were reached by the many brilliant physicists who worked in this field.

- Matter particles are called **fermions**. The fermions are divided into two groups: the **quarks** and the **leptons**.
- The leptons are very small. They include the electron as well as the **tau** particle, **muons** and **neutrinos**. We will learn more about the discovery of neutrinos later.
- There are six types of quarks: **up**, **down**, **charm**, **strange**, **top** and **bottom**. They are arranged in three 'generations'. In the current low-energy state of the Universe, only the lightest generation of quarks (the up and down quarks) exist for a significant length of time. Each generation of quarks is also associated with a pair of leptons. Table 9.1 shows each generation of quarks and their associated leptons.

Generation	Name	Symbol	Charge	Name	Symbol	Charge	Name	Symbol	Charge	Name	Symbol	Charge
I	Up	u	$+\frac{2}{3}$	Down	d	$-\frac{1}{3}$	Electron	e	−1	Electron neutrino	υ_e	0
II	Charm	c	$+\frac{2}{3}$	Strange	s	$-\frac{1}{3}$	Muon	μ	−1	Muon neutrino	υ_μ	0
III	Top	t	$+\frac{2}{3}$	Bottom	b	$-\frac{1}{3}$	Tau	τ	−1	Tau neutrino	υ_τ	0

Table 9.1 Matter quarks and their associated leptons

- Each fundamental particle of matter has an equivalent **antimatter** particle. One way of thinking about antimatter particles is that they are the same as the matter particle, but their charge is the opposite polarity. For example, the antiparticle of the electron is the **positron**. It has the same mass as an electron but it has a charge of +1. The antimatter quarks and leptons can be summarised in a table similar to that for matter particles (Table 9.2).

Generation	Name	Symbol	Charge	Name	Symbol	Charge	Name	Symbol	Charge	Name	Symbol	Charge
I	Anti-Up	u	$-\frac{2}{3}$	Anti-Down	d	$+\frac{1}{3}$	Positron	e⁺	+1	Anti-Electron neutrino	υ_e	0
II	Anti-Charm	c	$-\frac{2}{3}$	Anti-Strange	s	$+\frac{1}{3}$	Anti-Muon	$\bar{\mu}$	+1	Anti-Muon neutrino	υ_μ	0
III	Anti-Top	t	$-\frac{2}{3}$	Anti-Bottom	b	$+\frac{1}{3}$	Anti-Tau	$\bar{\tau}$	+1	Anti-Tau neutrino	υ_τ	0

Table 9.2 Antimatter quarks and leptons

- When a particle and its antiparticle combine they annihilate each other, releasing energy.
- The quarks are normally found in combination with each other. The particles that form when quarks combine are called **hadrons**. Two quarks combine together to form **mesons** and three quarks combine together to form **baryons**.
- Combinations of quarks can only exist when there is a whole number for the overall charge. For example, mesons comprise a combination of a quark and an antiquark. This means that these particles have very short lifetimes and are difficult to detect.
- The baryons contain two particles that we are familiar with: protons and neutrons. A proton is a combination of two up-quarks and one down-quark, giving a total charge of +1. A neutron is a combination of two down-quarks and one up-quark, giving a total charge of 0.
- Neutrinos are their own antiparticles.

Beta decay and neutrinos

When the **beta** decay of a nucleus was studied closely in experiments in the 1930s, it was discovered that the total energy before and after the decay did not match the expectations of the experimenters. This led the Italian physicist Enrico Fermi to propose that another particle was involved in the process. This particle would have a very small mass, no charge and would account for the missing energy. This particle is the **neutrino**. The first evidence for the existence of the neutrino comes from the study of beta decay.

The four fundamental forces and their mediating particles

There are four types of force in the Universe: the strong nuclear force, the weak nuclear force, the electromagnetic force and the gravitational force. Each of these forces has a **force-mediating particle** or particles associated with it. These particles are called the **gauge bosons**. When two massive particles (objects with mass) interact, they exchange a gauge boson. It is this exchange or transfer of particles that leads to forces being applied from one object to another.

The gravitational force is the force of attraction between all massive objects in the Universe. It is by far the weakest of the forces but it is infinite in range. Its force-mediating particle has never been detected but it has been given the name the **graviton**.

The electromagnetic force also acts over infinite distances. It is the force of everyday interactions between objects and its strength is very much greater than that of the gravitational force (around 36 orders of magnitude greater). The force-mediating particle for the electromagnetic force is the photon.

The other two fundamental forces of nature both have very short ranges and are only effective within the radius of the nucleus.

● The weak force is associated with beta decay and its gauge bosons are the **W⁺**, **W⁻** and **Z bosons**. This force is around 25 orders of magnitude greater than the gravitational force.

● The strong force is required to keep the particles that make up the nucleus together. Its gauge boson is the **gluon** and it has a strength 38 orders of magnitude greater than gravity.

The fundamental particles and forces can be summarised in the following diagram.

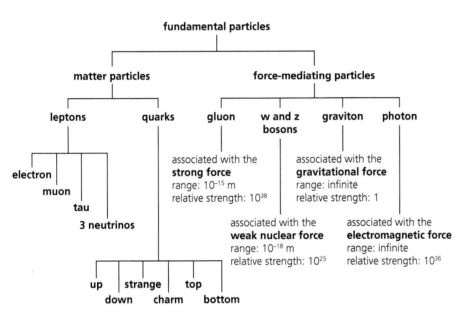

Figure 9.1 Fundamental particles and forces diagram

Key points

* You should understand orders of magnitude, from the very small to the very large, and be able to give examples within certain size ranges.
* There is evidence for the existence of subnuclear particles and antimatter.
* The Standard Model details the particles that make up the Universe.
* Fermions are the matter particles and are made up of two groups, quarks and leptons.
* Hadrons are made up of six types of quarks.
* Baryons are hadrons that contain three quarks; mesons contain two quarks.
* The leptons contain no smaller subdivisions; the electron is an example.
* The force-mediating particles are called bosons.
* The first evidence for one of the leptons, the neutrino, was found by analysing beta decay in atoms.

Key words

Antimatter – material made up of antiquarks and antileptons. Each matter particle has an antimatter particle equivalent with an opposite charge but identical mass

Baryon – a particle made up of three quarks

Boson – a force-mediating particle or a meson

Fermion – a matter particle

Hadron – a particle made up of quarks

Leptons – light particles that are themselves fundamental

Meson – a particle made up of two quarks

Neutrino – one of the leptons, first discovered by studying beta decay

Positron – the antiparticle of the electron

Quarks – the fundamental particles that make up the hadrons

Tau – one of the leptons

Questions

1　Which group of particles does the electron belong to?
2　How many types of quarks are there?
3　Where did the first evidence of neutrinos come from?
4　Which quarks make up a neutron?
5　Which quarks make up an antiproton?
6　Why are neutral mesons very short-lived?
7　Name the four forces of nature and their force-mediating particles.

Nuclear reactions

Nuclear symbols

The nucleus of an atom consists of protons and neutrons. The number of protons and neutrons in a nucleus can be represented by two numbers: the atomic number and the mass number.

 The atomic number gives the number of protons in the nucleus. This identifies the element – all atoms of a particular element have the same number of protons in the nucleus. We can use data booklets to identify elements given the atomic number.

 The mass number gives the total number of protons plus the total number of neutrons in the nucleus. This identifies the **isotope** of the element. Different isotopes of an element have the same number of protons but different numbers of neutrons. Different isotopes of an element have similar chemical properties to each other but undergo different nuclear reactions.

The mass number and the atomic number are represented in the following symbolic form:

$$^A_Z X$$

where

    X is the element symbol

    A is the mass number

    Z is the atomic number.

Nuclear decay

Some isotopes are unstable – they will undergo spontaneous nuclear decay. The types of radioactive decay are **alpha**, **beta** and **gamma**. All types of nuclear decay tend to make atoms more stable.

In alpha decay, a nucleus emits a particle with two protons and two neutrons, called an alpha (α) particle. As a result of this process, the atomic number of the isotope is reduced by two and the mass number by four.

For example:

$$^{238}_{92}U \rightarrow \,^{234}_{90}Th + \,^{4}_{2}\alpha.$$

In this example above, uranium has emitted an alpha particle to become thorium. Alpha particles are often represented as a helium nucleus.

Some isotopes decay by emitting a beta (β) particle. During beta decay, a neutron decays to become a proton. This results in a beta particle and a neutrino being emitted by the neutron. The beta particle is in fact an electron. This causes the atomic number of the isotope to increase by one but results in no change in the mass number. For example:

$$^{14}_{6}C \rightarrow \,^{14}_{7}N \,+ \,^{0}_{-1}\beta + \upsilon$$

Other isotopes emit a high-energy photon of electromagnetic radiation to become more stable. This is called gamma (γ) emission. Gamma photons are part of the electromagnetic spectrum and so have no mass or charge. Gamma emission does not affect the atomic number or the mass number of the isotope.

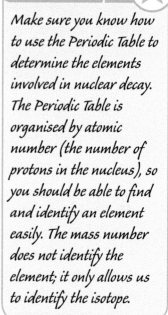

Hints & tips

Make sure you know how to use the Periodic Table to determine the elements involved in nuclear decay. The Periodic Table is organised by atomic number (the number of protons in the nucleus), so you should be able to find and identify an element easily. The mass number does not identify the element; it only allows us to identify the isotope.

Nuclear fission

When a larger nucleus splits into two smaller nuclei this is called **nuclear fission**. There are two types of nuclear fission: **spontaneous** and **induced**.

Spontaneous nuclear fission is a natural process in which a larger nucleus splits into two smaller nuclei. This process also results in the release of neutrons that will not be contained in the two smaller nuclei. The spontaneous decay of uranium-236 is a good example:

$$^{236}_{92}U \rightarrow \,^{144}_{56}Ba + \,^{89}_{36}Kr + 3\,^{1}_{0}n$$

The total of the mass numbers and atomic numbers on the left-hand side and the right-hand side of the equation must be equal. It is important that you can balance nuclear equations.

Induced fission takes place when a large nucleus is bombarded with neutrons and absorbs one of them. This makes the nucleus very unstable, causing it to split. For example:

$$^{235}_{92}U + \,^{1}_{0}n \rightarrow \,^{236}_{92}U \rightarrow \,^{144}_{56}Ba + \,^{89}_{36}Kr + 3\,^{1}_{0}n$$

It is important that you can identify and explain spontaneous and induced fission. To determine which type of reaction you are looking at, study the left-hand side of the statement. If there is a neutron present then the reaction is induced, if there is not the reaction is spontaneous. Induced fission statements need to be balanced in the same way as spontaneous fission statements.

The neutrons released in this fission reaction can go on to split other nuclei. This is the basis of nuclear **chain reactions** that are used to produce energy in nuclear reactors.

Nuclear fusion

Nuclear fusion takes place when two small nuclei join together to make a larger nucleus. This is the process that takes place in stars and releases energy. The reaction below shows how two nuclei of tritium, an isotope of hydrogen, combine to produce helium in a fusion reaction.

$$^{3}_{1}H + {}^{3}_{1}H \rightarrow {}^{4}_{2}He + 2{}^{1}_{0}n$$

As with fission reactions, the statement needs to be balanced.

The equivalence of mass and energy

In all types of nuclear reaction, the total mass on the left-hand side of the statement is greater than the total mass on the right-hand side. This is because some of the original mass has been converted into energy. Mass and energy are equivalent. The equivalence is given by Einstein's famous equation:

$$E = mc^2$$

We can use this equation to calculate the amount of energy released in each reaction.

 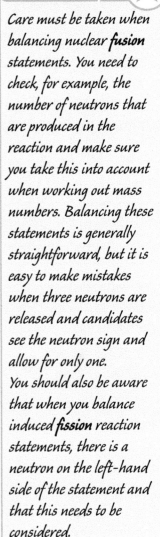
Example

When we examine the spontaneous fission of uranium in the reaction below we can calculate the amount of energy released. Table 10.1 gives the masses of the isotopes present in the reaction plus the mass of a neutron.

$$^{236}_{92}U \rightarrow {}^{144}_{56}Ba + {}^{89}_{36}Kr + 3{}^{1}_{0}n$$

Particle	Mass/kg
$^{236}_{92}U$	$3{\cdot}920 \times 10^{-25}$
$^{144}_{56}Ba$	$2{\cdot}390 \times 10^{-25}$
$^{89}_{36}Kr$	$1{\cdot}477 \times 10^{-25}$
$^{1}_{0}n$	$1{\cdot}675 \times 10^{-27}$

Table 10.1 Uranium fission reaction masses

The total mass on the left-hand side of the statement is the mass of the $^{236}_{92}U = 3{\cdot}920 \times 10^{-25}$ kg.

The total mass on the right-hand side is the mass of $^{144}_{56}Ba$ plus the mass of $^{89}_{36}Kr$ plus the mass of three neutrons:

$$2{\cdot}390 \times 10^{-25}$$
$$1{\cdot}477 \times 10^{-25}$$
$$+ \quad 3 \times 1{\cdot}675 \times 10^{-27} = 5{\cdot}025 \times 10^{-27} \quad = 0{\cdot}050 \times 10^{-25}$$
$$\overline{3{\cdot}917 \times 10^{-25}\text{ kg}}$$

$\Rightarrow$

The difference in mass between the left-hand side and the right-hand side is:

$$\begin{array}{r} 3{\cdot}920 \times 10^{-25} \\ - \quad 3{\cdot}917 \times 10^{-25} \\ \hline 0{\cdot}003 \times 10^{-25} \, \text{kg} \end{array}$$

The energy released is calculated using $E = mc^2$:

$E = 0{\cdot}003 \times 10^{-25} \times (3 \times 10^8)^2$

$E = 2{\cdot}7 \times 10^{-11} \, \text{J}$

It can be seen that this is a very small amount of energy. It is important to remember that this is the energy released when a single nucleus undergoes fission. In any fission reaction there are likely to be billions of atoms involved and so large amounts of energy will be released.

The energy released from induced nuclear fission and from fusion reactions is calculated in the same way. The fusion reaction involving tritium ions can be analysed as follows.

Example

$$^{3}_{1}\text{H} + ^{3}_{1}\text{H} \rightarrow ^{4}_{2}\text{He} + 2\,^{1}_{0}\text{n}$$

Particle	Mass/kg
$^{3}_{1}\text{H}$	$5{\cdot}006 \times 10^{-27}$
$^{4}_{2}\text{He}$	$6{\cdot}642 \times 10^{-27}$
$^{1}_{0}\text{n}$	$1{\cdot}675 \times 10^{-27}$

Table 10.2 Hydrogen fusion reaction masses

The mass of the left-hand side is $2 \times 5{\cdot}006 \times 10^{-27} = 10{\cdot}012 \times 10^{-27} \, \text{kg}$.

The mass of the right-hand side is the mass of the helium nucleus plus the mass of two neutrons:

$$\begin{array}{r} 6{\cdot}642 \times 10^{-27} \\ + \quad 2 \times 1{\cdot}675 \times 10^{-27} = 3{\cdot}350 \times 10^{-27} \\ \hline 9{\cdot}992 \times 10^{-27} \, \text{kg} \end{array}$$

The difference in mass between the left-hand side and the right-hand side is:

$$\begin{array}{r} 10{\cdot}012 \times 10^{-27} \\ - \quad 9{\cdot}992 \times 10^{-27} \\ \hline 0{\cdot}020 \times 10^{-27} \, \text{kg} \end{array}$$

Using $E = mc^2$, the energy released is therefore:

$E = 0{\cdot}02 \times 10^{-27} \times (3 \times 10^8)^2$

$E = 1{\cdot}8 \times 10^{-12} \, \text{J}$

This is an even smaller amount of energy than for the fission reaction. It must be remembered again that this is the energy released in the fission of two individual hydrogen nuclei to form one nucleus of helium.

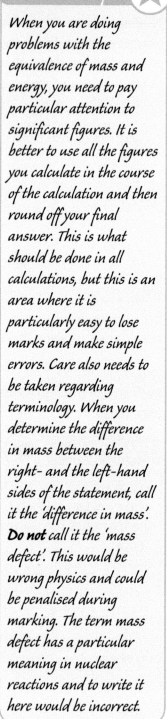

Hints & tips

When you are doing problems with the equivalence of mass and energy, you need to pay particular attention to significant figures. It is better to use all the figures you calculate in the course of the calculation and then round off your final answer. This is what should be done in all calculations, but this is an area where it is particularly easy to lose marks and make simple errors. Care also needs to be taken regarding terminology. When you determine the difference in mass between the right- and the left-hand sides of the statement, call it the 'difference in mass'. **Do not** call it the 'mass defect'. This would be wrong physics and could be penalised during marking. The term mass defect has a particular meaning in nuclear reactions and to write it here would be incorrect.

Fusion reactors

Nuclear fusion is a promising technology that may be used in the future to produce energy commercially for homes and industry. At the moment, however, there are problems with several aspects of building a fusion reactor.

In order for the fusion reactions to take place, very high temperatures are required (about 10 million kelvin). This produces a plasma that needs to be contained inside the reactor vessel. This is done using a very powerful magnetic field produced by electromagnets. A great deal of power is required to maintain the electromagnets and is currently one of the obstacles to producing a commercially viable fusion reactor.

There are also problems resulting from the extremely high temperature of the plasma, as it is hot enough to melt or evaporate the walls of the reactor.

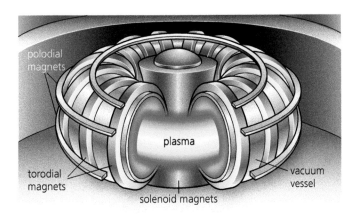

polodial magnets

torodial magnets

plasma

solenoid magnets

vacuum vessel

Figure 10.1 A nuclear fusion reactor

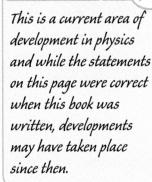

Hints & tips

This is a current area of development in physics and while the statements on this page were correct when this book was written, developments may have taken place since then.

Once a reactor is up and running, there are further issues with extracting energy from the plasma as the reactor continues to run. If these difficulties can be overcome then fusion reactors may provide a long-term solution to our energy needs.

Key points

* Nuclear statements can be written to describe nuclear decay, fission reactions and fusion reactions.
* Nuclear isotopes can be represented in symbol form.
* The Periodic Table can be used to identify isotopes from their symbol.
* There are three types of variation in the decay of radioactive isotopes: alpha, beta and gamma.
* You should be able to tell the difference between spontaneous and induced nuclear fission.
* Calculations can be carried out involving the equivalence of mass and energy.
* There are issues surrounding cooling and containment in nuclear fusion reactors which must be understood.

Key words

Alpha – a type of radioactive decay; when an isotope undergoes alpha decay, it releases a particle made up of two protons and two neutrons

Beta – a type of radioactive decay; during beta decay a neutron in an isotope decays to become a proton, releasing a beta particle (an electron) and a neutrino

Fission – the splitting of a nucleus into two smaller nuclei

Fusion – the joining of two small nuclei to form a larger nucleus

Gamma – photons of electromagnetic energy released by a nucleus

Induced – caused to happen; in the case of nuclear fission the addition of a neutron to a large nucleus, causing it to split into two smaller nuclei

Isotope – isotopes of an element have the same number of protons but different numbers of neutrons; these isotopes have identical chemical properties but different nuclear reactions

Spontaneous – happens naturally; in nuclear fission an isotope that decays without being bombarded by neutrons

Questions

The following statement will use data from Table 10.3.

Particle	Mass/kg
$^{2}_{1}H$	3.342×10^{-27}
$^{3}_{1}H$	5.005×10^{-27}
$^{4}_{2}He$	6.642×10^{-27}
$^{95}_{42}Mo$	157.544×10^{-27}
$^{139}_{57}La$	230.584×10^{-27}
$^{235}_{92}U$	390.173×10^{-27}
$^{0}_{1}n$	1.675×10^{-27}

Table 10.3 Particle and mass information

1 The following statement describes the nuclear decay of an isotope of lead:

$$^{214}_{82}Pb \rightarrow {}^{214}_{83}Bi + {}^{A}_{B}X + \upsilon$$

a) Calculate values for A and B and hence identify the type of decay taking place.

b) Name particle υ.

2 Figure 10.2 shows part of a decay series.

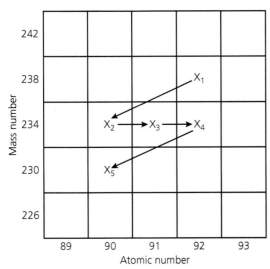

Figure 10.2 Graph showing part of a decay series

a) Use a Periodic Table to identify elements X_1 and X_5.
b) Calculate the number of alpha particles emitted in decaying from X_1 to X_5.
c) Calculate the number of beta particles emitted in decaying from X_1 to X_5.

3 The following nuclear reaction takes place:

$$^{7}_{3}\text{Li} + ^{2}_{1}\text{H} \rightarrow 2^{4}_{2}\text{He} + ^{1}_{0}\text{n}$$

During this reaction, $2 \cdot 97 \times 10^{-12}$ J of energy are released. Calculate the mass of the $^{7}_{3}$Li isotope.

4 The following fission reaction takes place:

$$^{235}_{92}\text{U} + ^{1}_{0}\text{n} \rightarrow ^{139}_{57}\text{La} + ^{95}_{42}\text{Mo} + 2^{1}_{0}\text{n} + 7^{0}_{-1}\text{e}$$

a) State whether this reaction is spontaneous or induced.
b) Calculate the energy released in the reaction. (The mass of the $-^{0}_{1}$e particles are negligible.)

*5 The Sun is the source of most of the energy on Earth. This energy is produced by nuclear reactions which take place in the interior of the Sun. One such reaction can be described by the following statement:

$$^{3}_{1}\text{H} + ^{2}_{1}\text{H} \rightarrow ^{4}_{2}\text{He} + ^{1}_{0}\text{n}$$

The masses of the particles involved in this reaction are shown in the table below.

Particle	Mass/kg
$^{3}_{1}$H	$5 \cdot 005 \times 10^{-27}$
$^{2}_{1}$H	$3 \cdot 342 \times 10^{-27}$
$^{4}_{2}$He	$6 \cdot 642 \times 10^{-27}$
$^{1}_{0}$n	$1 \cdot 675 \times 10^{-27}$

Table 10.4 Particle and mass information

a) Name this type of nuclear reaction.
b) Calculate the energy released in this reaction.

6 Which of the following statements describes an induced nuclear reaction?

A $\ ^{235}_{92}U + \ ^{1}_{0}n \rightarrow \ ^{144}_{56}Ba + \ ^{90}_{36}Kr + 2\ ^{1}_{0}n$

B $\ ^{7}_{3}Li + \ ^{1}_{1}H \rightarrow \ ^{4}_{2}He + \ ^{4}_{2}He$

C $\ ^{3}_{1}H + \ ^{2}_{1}H \rightarrow \ ^{4}_{2}He + \ ^{1}_{0}n$

D $\ ^{226}_{88}Ra \rightarrow \ ^{222}_{86}Rn + \ ^{4}_{2}He$

E $\ ^{216}_{84}Po \rightarrow \ ^{216}_{84}Po + \gamma$

*7 The statement below represents a nuclear reaction.

$$\ ^{3}_{1}H + \ ^{2}_{1}H \rightarrow \ ^{4}_{2}He + \ ^{1}_{0}n$$

The total mass on the left-hand side is 8.347×10^{-27} kg.
The total mass on the right-hand side is 8.316×10^{-27} kg.
The energy released during one nuclear reaction of this type is:

A 9.30×10^{-21} J

B 2.79×10^{-12} J

C 7.51×10^{-10} J

D 1.50×10^{-9} J

E 2.79×10^{15} J

Wave properties

What you should know

★ Conditions for constructive and destructive interference
★ Coherent waves have a constant phase relationship and have the same frequency, wavelength and velocity
★ Constructive and destructive interference in terms of phase between two waves
★ Interference of waves using two coherent sources
★ Maxima and minima are produced when the path difference between waves is a whole number of wavelengths or an odd number of half wavelengths, respectively
★ The relationship between the wavelength, distance between the sources, distance from the sources and the spacing between maxima or minima

Hints & tips

*This section of the course builds on knowledge that you should have gained when studying National 5 Physics. It is assumed that students will be familiar with the general properties of a wave: wavelength, amplitude, etc. In addition, a limited knowledge of **diffraction** and the basic wave relationships have been taken as familiar. You should also be aware that wave properties are often discussed using a particular type of wave as an example, but that the wave property applies to all waves. Interference, for example, is often discussed in terms of sound or light but you should be aware that all waves experience interference, so do not be surprised by an interference problem on microwaves or radio waves.*

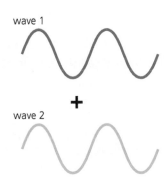

Figure 11.1 Waves in phase

Phase

Phase is an important concept used to describe the relationship between two or more waves. If two waves meet each other and are in phase, this means that a crest is meeting a crest or a trough is meeting a trough. If two waves meet each other but are completely out of phase then a crest will meet a trough.

When we look at how the interference of waves occurs, the phase relationship between two waves will be important.

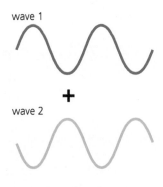

Figure 11.2 Waves out of phase

Coherence

Coherence is an important idea in the study of waves. Two waves are said to be coherent if they have the same speed, frequency, wavelength and have a constant phase relationship.

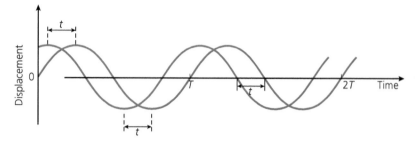

Figure 11.3 Coherent waves

The two waves in Figure 11.3 are described as being coherent. They have a constant phase relationship because the time, t, between corresponding points on the two waves is the same.

Interference

Interference is often described as the determining test for a wave. It is important to be precise about the language we use when describing something as being a wave or a particle. This is something that will be covered in greater depth in Chapter 12 when we discuss wave–particle duality. For now, we will say that it would be better to describe interference as the test for a wave nature. It is essentially strong evidence for the use of a wave model to explain the properties of light.

Interference can be demonstrated with a variety of different sources in a laboratory such as sound waves, light waves and microwaves. When two speakers are set up as shown in Figure 11.4, an interference pattern is produced.

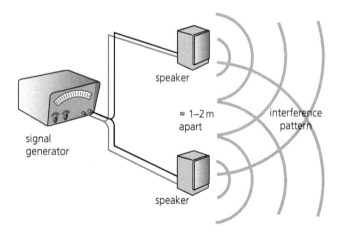

Figure 11.4 Demonstrating interference with sound waves

If you walked across the room in front of the two speakers, you would notice areas where the sound was loud and areas where the sound was very quiet or silent. This is because the two speakers are acting as coherent sources of sound waves and producing an interference pattern.

The loud regions are zones of **constructive interference** (**maxima**) and the quieter regions are zones of **destructive interference** (**minima**).

At maxima, two waves that are in phase with each other combine to produce a louder sound than just one wave. At minima, two waves that are exactly out of phase meet and cancel each other out. This produces points of silence in this pattern.

It is important to note that waves produce interference patterns but particles do not. This means that if an interference pattern is observed it is evidence of a wave nature.

An interference pattern can be produced in a water tank by having two circular dippers produce circular coherent waves on the surface. The pattern typically produced is shown in Figure 11.5 as both a photograph and a line drawing.

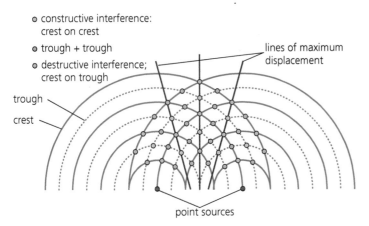
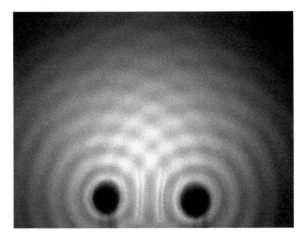

Figure 11.5 An interference pattern in a water tank

We can see that in the pattern produced, both the areas of constructive interference and the areas of destructive interference are straight lines.

Path difference

When we get a point of constructive interference, we can think of this as a crest from one source combining with a crest from another source. If the waves are coherent then the two sources are in phase with each other. This means that when a crest from one source meets a crest from the other source, either the waves must have travelled the same distance or one wave must have travelled a whole number of wavelengths more than the other.

At a point of destructive interference, a crest meets a trough. When the sources are coherent this must mean that one wave has travelled half a wavelength, or one and a half wavelengths, or two and a half wavelengths, etc. more than the other.

The difference in the distance from one source to a point in the interference pattern and the distance from the other source to the point is called the **path difference**. When the path difference is zero or a whole number of wavelengths then the point is a maximum (or a point of constructive interference). If the path difference is an odd number of half wavelengths ($\frac{1}{2}$, $\frac{3}{2}$, $\frac{5}{2}$, etc.) then the point is a minimum (or a point of destructive interference).

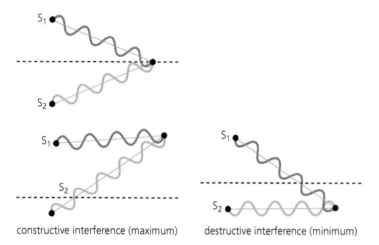

constructive interference (maximum)   destructive interference (minimum)

Figure 11.6 Waves combining and interfering

This can be expressed mathematically:
- for a maximum $S_2P - S_1P = m\lambda$
- for a minimum $S_2P - S_1P = (m + \frac{1}{2})\lambda$

where

S$_1$P is the distance from source one to the point in the interference pattern

S$_2$P is the distance from source two to the point in the interference pattern

m is a whole number (0, 1, 2 …)

λ is the wavelength of the waves.

For the central maximum in an interference pattern, $m = 0$. This is because the distance from both sources to a central maximum must be the same, so the path difference is zero.

If we look at Figure 11.7 which shows microwave interference, we can work out the wavelength of the microwaves.

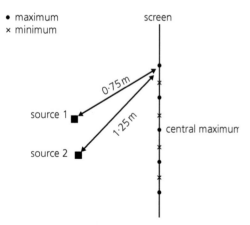

Figure 11.7 Microwave interference

As we count from the central maximum to the point we are analysing, we see that:

$m = 2$, $S_2P = 1·25$ m and $S_1P = 0·75$ m.

Now $S_2P - S_1P = m\lambda$

So $1·25 - 0·75 = 2\lambda$

$$0·50 = 2\lambda$$

$$\lambda = 0·25 \text{ m}$$

From the same diagram we can also determine the path difference to the minimum 'below' the maximum we have just calculated. For this point the path difference is $1\frac{1}{2}\lambda$.

Thus $S_2P - S_1P = 1\frac{1}{2}\lambda$

$$S_2P - S_1P = 1\frac{1}{2} \times 0·25$$

$$S_2P - S_1P = 0·375 \text{ m}$$

So the path difference for this point is $0·375$ m.

Hints & tips ★

Many students find path difference problems difficult initially. Remember that the path is a distance.

S_2P is the distance from source 2, S_2, to the point P.

S_1P is the distance from source 1, S_1, to the point P.

So $S_2P - S_1P$ is just the difference between those two distances.

Gratings

When plane waves pass through a gap that is narrower than one wavelength, the waves that emerge have a circular wavefront.

Figure 11.8 This is due to the diffraction of the waves.

It is possible to produce an interference pattern with light using a **grating**. A grating is a piece of glass with very fine lines etched onto it. The gaps between the lines act as slits that produce coherent beams of light with circular wavefronts. These beams interfere with each other to produce an interference pattern.

Figure 11.9 shows the effect produced when a laser is used with a grating. The bright spots are points of constructive interference (maxima) and the dark areas are points of destructive interference (minima).

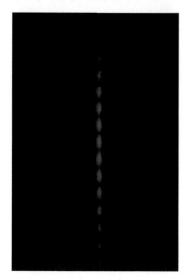

Figure 11.9 The interference pattern caused when a laser is shone through a grating

The mathematical relationship for an interference pattern produced from a grating is

$$m\lambda = d \sin \theta$$

where

m is the order of the maximum, that is, the first maximum, second maximum, etc.

λ is the wavelength of the light (or electromagnetic radiation)

d is the separation of the lines/slits on the grating

θ is the angle as shown in Figure 11.10.

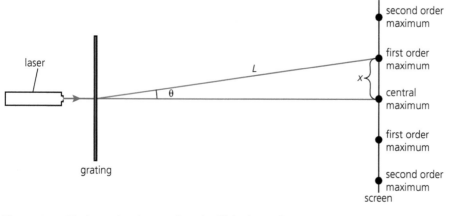

Figure 11.10 To determine the wavelength of light from a laser

Figure 11.11 shows the interference pattern produced by a laser when the slit separation is 3.50×10^{-6} m.

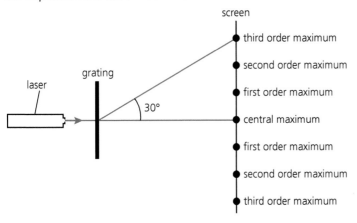

Figure 11.11 Interference pattern from a laser

Example

We can use the grating equation to calculate the wavelength of the laser light.

Since $m\lambda = d \sin \theta$

$3 \times \lambda = 3.50 \times 10^{-6} \times \sin 30°$

$\lambda = (3.50 \times 10^{-6} \times 0.5) \div 3$

$\lambda = 580$ nm

Key points !

* Coherent waves have the same speed, wavelength and frequency as well as a constant phase relationship.
* The test for a wave nature is interference.
* Constructive interference takes place when two waves that are in phase meet.
* Destructive interference takes place when two waves that are exactly out of phase meet.
* In an interference pattern, the path difference is the difference in distance from one source to a point in the interference pattern and from the other source to the point.
* Constructive interference (maxima) takes place when the path difference is zero or a whole number of wavelengths.
* Destructive interference (minima) takes place when the path difference is an odd number of half wavelengths.
* You should be able to solve problems using the path difference equations:

$$S_2P - S_1P = m\lambda \text{ and } S_2P - S_1P = (m + \tfrac{1}{2})\lambda$$

* When waves pass an edge or through a narrow gap or aperture, diffraction takes place.
* If the aperture is smaller than the wavelength of the waves, the waves passing through the aperture will have circular wavefronts.
* You should be able to solve problems using the grating equation: $m\lambda = d \sin \theta$

Key words

Coherent – waves that are coherent have the same speed, wavelength and frequency and have a constant phase relationship

Constructive – two waves combining to produce a maximum

Destructive – two waves combining to produce a minimum

Diffraction – the spreading of waves on passing an edge or through a narrow aperture

Grating – a sheet of transparent material with fine lines etched on it to produce diffraction; each gap in the grating acts as a coherent source

Interference – a series of maxima and minima produced when waves meet

Maxima – points of constructive interference

Minima – points of destructive interference

Path difference – the difference in distance from one source to a point and the other source to the point

Phase – how we describe the relative position of a wave

Questions

1 What is meant by the term 'coherent waves'?
2 Explain, using the superposition of waves, how constructive interference takes place.
3 A sound interference pattern is produced using two speakers that act as coherent sources of sound.

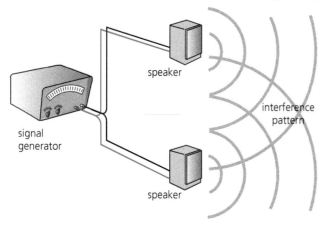

Figure 11.12 Interference with sound waves

A student stands at a maximum that is the same distance from both speakers. The student then moves to the next maximum in the pattern. At this point she is 0·75 m from one speaker and 1·43 m from the other speaker.

a) Calculate the wavelength of the sound waves.

b) The speed of sound in air is 340 m s⁻¹. Calculate the frequency of the signal generator.

4 Monochromatic light from a laser is shone onto a grating with 400 lines per millimetre. The angle between the central bright fringe and the second order maximum is 25°. Calculate the wavelength of the laser light.

5 Two identical loudspeakers L$_1$ and L$_2$ are connected to a signal generator as shown.

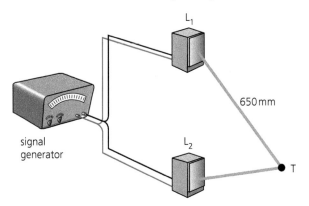

Figure 11.13 Two loudspeakers connected to a signal generator

An interference pattern is produced. A minimum is detected at point T. The wavelength of the sound is 30 mm. The distance from L$_1$ to T is 650 mm. The distance from L$_2$ to T is:

A 605 mm

B 615 mm

C 625 mm

D 645 mm

E 655 mm.

6 S$_1$ and S$_2$ are sources of coherent waves. An interference pattern is obtained between X and Y.

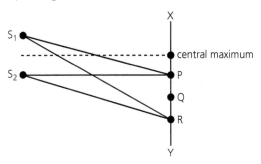

Figure 11.14 Two sources of coherent waves

The first order maximum occurs at P, where S$_1$P = 210 mm and S$_2$P = 235 mm. The third order maximum occurs at R. The path difference at R is:

A 25 mm

B 50 mm

C 75 mm

D 100 mm

E 125 mm.

Chapter 12
Spectra

> ### What you should know
>
> ★ The photoelectric effect as evidence for the particulate nature of light
> ★ Photons of sufficient energy can eject electrons from the surface of materials
> ★ The threshold frequency is the minimum frequency of a photon required for photoemission
> ★ The work function of the material is the minimum energy required to cause photoemission
> ★ Determination of the maximum kinetic energy of photoelectrons
> ★ Irradiance and the inverse square law
> ★ Irradiance is power per unit area
> ★ The relationship between irradiance and distance from a point light source
> ★ Line and continuous emission spectra, absorption spectra and energy-level transitions
> ★ The Bohr model of the atom
> ★ Movement of electrons between energy levels
> ★ The terms ground state, energy levels, ionisation and **zero potential energy** for the Bohr model of the atom
> ★ Emission of photons due to movement of electrons between energy levels and dependence of photon frequency on energy difference between levels
> ★ The relationship between photon energy, Planck's constant and photon frequency
> ★ Absorption lines in the spectrum of sunlight provide evidence for the composition of the Sun's upper atmosphere

Wave–particle duality

The **photoelectric effect** is the name given to phenomena where light or electromagnetic radiation can cause electrons to move or be transferred from one material to another.

In the late 1800s, light was thought of as a continuous waveform where the amount of energy transferred could be determined by the intensity and duration of the light. A number of experiments at this time, however, indicated that ultraviolet light was causing ionisation of gases. A number of theories were proposed but Albert Einstein's description and explanation (1905) was found to explain in some detail all the observed experimental results. He finally received the Nobel Prize for Physics for his work in 1921.

The experimental results were:

- a negatively charged metal disc could be discharged by shining UV light upon it
- shining more 'intense' light of a lower frequency had no effect; it would not discharge.

Einstein's explanation was that light behaved like, or exhibited some characteristics of, a particle. He proposed that light travelled in small amounts, **quanta** or **photons**, and went on to suggest that the energy that a photon has is related to its frequency. A higher-frequency photon has more energy associated with it than one of a lower frequency. Indeed, a high-frequency photon from ultraviolet light has enough energy to eject an electron from a metal. The lower-frequency photons do not and, while they can bombard the electron more often, they do not have enough energy as a 'bundle' to eject the electron. Orange light, for example, would never be able to eject an electron from a disc of metal.

This was a major change in the understanding of physics. It showed that in certain situations light behaved as a particle, which went against the belief that light was a continuous wave. This revolutionised our understanding of physics at this point and led to the beginning of quantum mechanics, which describes the interactions of energy and matter at the molecular level.

It was found that for a particular material, only photons above a certain frequency will be able to eject electrons. This is known as the **threshold frequency** (f_o) and it differs for different materials.

The energy, E, of a particular photon can be determined by its frequency using the relationship:

$$E = hf$$

where f is its frequency (in Hz) and h is **Planck's constant** ($6.63 \times 10^{-34}\,\mathrm{J\,s}$).

Example

1 Calculate the energy associated with photons of the following frequency
   a) $f = 5.3 \times 10^{15}\,\mathrm{Hz}$

   b) $f = 7.2 \times 10^{17}\,\mathrm{Hz}$.

2 Light of wavelength 650 nm is incident upon a metal. Calculate the energy associated with this wavelength.

Solution

1 **a)** $E = hf$
    $\qquad = 6.63 \times 10^{-34} \times 5.3 \times 10^{15} = 3.5 \times 10^{-18}\,\mathrm{J}$

   b) $E = hf$
    $\qquad = 6.63 \times 10^{-34} \times 7.2 \times 10^{17} = 4.8 \times 10^{-16}\,\mathrm{J}$

2 $v = f\lambda$
    $\quad f = \dfrac{v}{\lambda} = 3 \times \dfrac{10^8}{650} \times 10^{-9} = 4.615 \times 10^{14}\,\mathrm{Hz}$
    $E = hf = 6.63 \times 10^{-34} \times 4.615 \times 10^{14} = 3.06 \times 10^{-19}\,\mathrm{J}$

The threshold frequency is the minimum frequency of radiation which can eject an electron from a material. This allows us to calculate the minimum energy required to eject an electron from the surface of a material using the equation $E = hf_o$. This is known as the **work function** for that material. If the energy associated with a photon is greater than the work function for that material, an electron will be ejected.

Consider a photon of frequency 7.82×10^{14} Hz when incident upon a material with a work function of 4.67×10^{-19} J.

The energy of the photon is given by $E = hf = 6.63 \times 10^{-34} \times 7.82 \times 10^{14}$ $= 5.18 \times 10^{-19}$ J.

The difference in energy is given by $5.18 \times 10^{-19} - 4.67 \times 10^{-19} = 0.51 \times 10^{-19}$ J.

This surplus energy is transferred to the electron and would be transformed to the kinetic energy of that electron.

In algebraic terms, $E = hf$. The term hf may be broken down into the equation for the photoelectric effect:

$$hf = hf_o + E_k$$

where

$\quad E$ = energy in joules

$\quad h$ = Planck's constant

$\quad f$ = frequency of incident photon

$\quad f_o$ = threshold frequency

$\quad E_k$ = the maximum kinetic energy of the ejected electron.

Modern devices

A number of modern devices rely upon the photoelectric effect in order to operate.

Photomultipliers are used as very sensitive detectors of light/photons around the visible spectrum. They operate on the principle that a photon strikes a sensitive detector. Generally these detectors are materials with a very low work function and they will eject an electron when the photon strikes the material. This electron is then accelerated and used in an amplification circuit, which increases the current by a factor of hundreds of thousands, allowing even individual photons to be intensified. Charge-coupled devices (CCD) such as these are used in night cameras and also in normal camera sensors. This technology allows photographs to be taken in poor light conditions which would have been very difficult a few years ago.

Irradiance

The **irradiance** at a surface is defined as the power per unit area. It is the amount of energy $1\,m^2$ of a surface receives every second from a source. It is important to realise that while irradiance normally refers to light, it can also be used to measure other waves such as microwaves, radio waves and gamma rays.

The formula for calculating irradiance is:

$$I = \frac{P}{A}$$

where

I is the irradiance in watts per square metre ($W\,m^{-2}$)

P is the power in watts (W)

A is the area in square metres (m^2).

This equation can be used to calculate the irradiance at any surface.

Lasers have a very high irradiance. The beam you see produced by a laser does not spread out in the way light from a bulb does. A normal light bulb radiates energy in three dimensions in order to light up a room and, as a result, the irradiance at a point in that room is very small. A laser 'spot' does spread with distance but not by a large amount; as a result, its irradiance remains high over a large distance.

An experiment can be carried out to measure how the irradiance varies with distance from a **point source** of light. A point source is an ideal source of light in which all of the light comes from a single point. In practice we use a lamp with a very small filament as our source and describe it as acting as a point source of light (Figure 12.1)

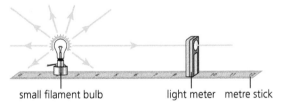

small filament bulb   light meter   metre stick

Figure 12.1 Ray of light passing through a light meter

The distance from the lamp is measured using a metre stick and the irradiance at that distance is measured using a light meter. This experiment is carried out in a darkened room. The results shown in Table 12.1 are typical of the data collected from the experiment.

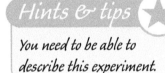
Distance/m	Irradiance/$W\,m^{-2}$
0·1	20·00
0·2	5·00
0·3	2·22
0·4	1·25
0·5	0·80
0·6	0·56
0·7	0·41
0·8	0·31
0·9	0·25
1·0	0·20

Table 12.1 Data for irradiance versus distance

When we plot these results we get the graph shown in Figure 12.2.

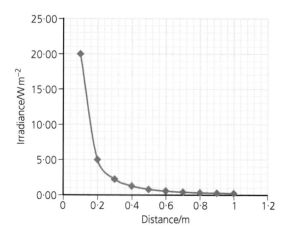

Figure 12.2 A graph of irradiance against distance

We can see that these data have produced a curve. In order to be able to write an equation for a relationship, we need a graph that is a straight line through the origin. Using these same results, we find that if we plot a graph of irradiance against $\frac{1}{d^2}$, we can attain this straight line relationship (Figure 12.3).

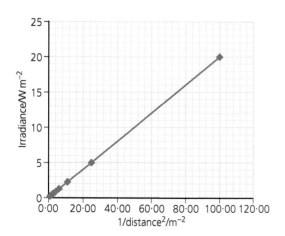

Figure 12.3 A graph of irradiance against 1/distance²

This allows us to state that:

$$I \propto \frac{1}{d^2}$$

So for any point light source:

$$I = \frac{k}{d^2}$$

where k is a constant.

We can see why this relationship comes about if we think about a beam of light from a point source spreading out with distance (see Figure 12.4).

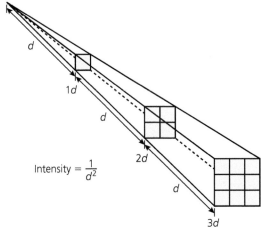

Figure 12.4 Light radiating from a point source

The total energy of the beam is the same all the way along its length but, as the distance from the source increases, the area the beam covers increases. At a distance, d, from the source, the area is one box but at double the distance, $2d$, the area is four boxes. This means that the irradiance at $2d$ must be quarter what it is at d. This so-called **inverse square relationship** is very common in physics and there are a number of physics relationships which are of this form.

Bohr's model of the atom

Early in the twentieth century, Niels Bohr developed a model of the atom that was able to explain many observations that had been made. This model was a refinement of earlier models of the atom. The development of the theory of the atom is interesting and worth further study, but we do not have room in this text to explore this topic.

In **Bohr's model**, an atom has a central positively charged nucleus with electrons in orbits around the nucleus. One important feature of this model is that the electrons are only allowed in certain fixed orbits and not in the spaces in-between.

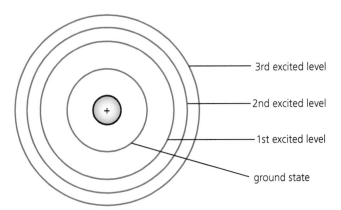

Figure 12.5 A simple model of an atom showing the central nucleus and the orbits or energy levels around it.

The electron orbits are often thought of as **energy levels** within the atom. The lowest level is referred to as the **'ground' state** and the higher energy levels as 'excited' states.

Hints & tips

The inverse square relationship is true for all electromagnetic radiations, not just visible light. This can be important, for example, when answering questions about gamma rays. If gamma rays are radiating away from a point source, then the inverse square relationship should be used. If you double your distance from the source, the irradiance will be one-quarter and so the dose of radiation received will be greatly reduced.

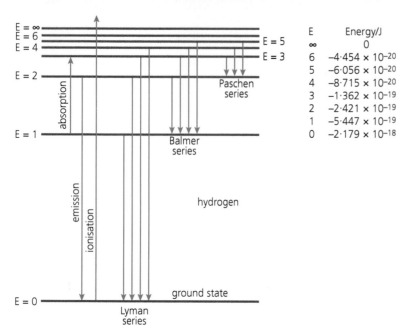

E	Energy/J
∞	0
6	$-4{\cdot}454 \times 10^{-20}$
5	$-6{\cdot}056 \times 10^{-20}$
4	$-8{\cdot}715 \times 10^{-20}$
3	$-1{\cdot}362 \times 10^{-19}$
2	$-2{\cdot}421 \times 10^{-19}$
1	$-5{\cdot}447 \times 10^{-19}$
0	$-2{\cdot}179 \times 10^{-18}$

Figure 12.6 The energy levels of electrons in a hydrogen atom

As electrons can only exist at certain energy levels, it follows that they can only move from one level to another by absorbing or releasing an exact quantity of energy – the difference in energy between the two levels. The electrical potential energy values for the energy levels in an atom are given as negative numbers (as shown in Figure 12.6). This is because they represent the electrical potential energy of an electron in the electric field of the nucleus.

If an electron is given enough energy, it will be removed from the atom altogether. This is called the **ionisation energy**. At this point the electrical potential energy of the electron in the nucleus' electric field is zero.

It should be noted that the Bohr model of the atom is only really effective for explaining the behaviour of the hydrogen atom, but it represented a major breakthrough in understanding the behaviour of all atoms.

Spectra

Emission spectrum

When we examine the spectrum of an element we see the following pattern.

Figure 12.7 The visible spectrum for hydrogen

We can now explain why we get this line spectrum from a particular element. All the atoms in an element are identical and will have identical energy levels. When an electron falls from one energy level to a lower one, it releases a fixed quantity of energy. This results in a photon of light with a certain energy being released. Each time an electron makes a transition between these two energy levels in an atom, a photon is released. Each one of these photons is identical so each has the same frequency. This produces a single line of a single colour. Each line in the spectrum is associated with a particular electron transition between two energy levels within the atom.

This type of spectrum is called the **emission spectrum** for that element. Each element has a different arrangement of energy levels within its structure so each element produces a different emission spectrum. This can be used to identify elements.

Absorption spectrum

If a white line is shone through the vapour of an element, it can be seen that the resulting spectrum has lines missing. This is called an **absorption spectrum**.

Figure 12.8 The absorption spectrum for hydrogen

These lines appear in exactly the same position as the emission lines for the element. This happens as the atom can absorb photons with exactly the same energy as the gap in the energy levels in the atom. The electrons absorb the energy from these photons and use it to move to a higher energy level in the atom. If a photon has an energy level other than the exact energy for a gap, it cannot be absorbed and so passes through the vapour. Each element absorbs a distinct pattern of lines (identical to the emission spectrum for the element). This process can be used to identify the elements present in a vapour.

When the spectrum of light from the Sun is observed, there are found to be dark lines in the continuous spectrum. These lines can be used by scientists to identify the elements present in the outer atmosphere of the Sun. They are known as Fraunhofer lines, after the German physicist who first identified them. They are caused by the elements present in the Sun's atmosphere absorbing photons from the Sun. There are over 500 of them.

Frequency and photon energy

The higher the energy gap in the atom, the higher the frequency of the photon produced. It was discovered that the two quantities are proportional to each other and so:

$E \alpha f$ or $E = hf$

where h is a constant known as Plank's constant. It has the value 6.63×10^{-34} J s.

This equation can be used to calculate the frequency of the emitted photon given the energy gap between two energy levels in an atom.

> ### Hints & tips ⭐
>
> *It is important that you can make the link between the Bohr model of the atom and the production of **line spectra**. Each line in a line spectrum corresponds to an electron transition inside the atom. The energy of the photon produced is equal to the energy transition in the atom.*

> ### Hints & tips ⭐
>
> *The larger the energy transition, the higher the frequency of the photon **but** the smaller the energy transition, the longer the wavelength of the photon.*

Hints & tips

Occasionally in an examination you will be asked to calculate the wavelength of a photon rather than the frequency. You can do this by first calculating the frequency using E = hf and then using v = f λ. The two equations can be combined to give $\lambda = \dfrac{hc}{E}$ where

    λ is the wavelength of the photon
    h is Plank's constant
    c is the speed of light
    E is the energy of the photon (which equals the energy of the electron transition).

This equation is not on the data sheet but can be useful.

Key points

* Photons of high frequency have a high energy associated with them.
* Photons with sufficient energy can eject electrons from materials.
* Photons providing energy in discrete bundles is evidence for light acting as particles.
* The energy required to eject an electron is known as the work function.
* The irradiance at a surface is the power per unit area and is measured in W m^{-2}.
* The equation $I = \dfrac{P}{A}$ can be used to calculate the irradiance at a surface.
* If you double the distance between yourself and a source, the irradiance falls to a quarter of its original value.
* Irradiance is inversely proportional to the square of the distance.
* The inverse square relationship for irradiance and distance applies to all electromagnetic radiations, not just light.
* You should understand and be able to describe the Bohr model of the atom.
* Electrons are only stable in certain energy levels in an atom.
* When electrons make the transition from a high energy level to a lower level, a photon of energy is released. The energy of the emitted photon is the same as the energy gap between the two energy levels.
* Atoms can absorb photons with certain energies. The energy of the absorbed photon is the same as the energy difference between two levels in the atom.
* For the Bohr model of the atom, an understanding of the terms: ground state, energy level, ionisation and zero potential energy.
* You should be able to use the transition of electrons between energy levels in an atom to explain line spectra and absorption spectra.
* You should understand the relationship between the energy of a photon and its frequency.
* You should be able to solve problems using $E = hf$.
* The presence of absorption lines in an analysis of the Sun's spectrum allows the identification of the elements present in the Sun's atmosphere.

Key words

Absorption spectrum – black lines in a continuous spectrum caused when the vapour of an element absorbs particular photons of light. The photons absorbed have exactly the same energy as energy gaps in the energy level within the atom of the element

Bohr model – a model of the atom with a central positive nucleus and electrons in definite energy levels surrounding the nucleus

Emission spectrum – lines of colour in a continuous spectrum associated with a particular electron transition between two energy levels within an atom

Energy level – an electron orbit inside the Bohr atom

Ground state – the lowest permitted energy level in an atom

Ionisation energy – the energy level at which an electron escapes from the electric field of the nucleus

Irradiance – the power per unit area of a radiation

Line spectra – lines with certain frequencies emitted by an element

Photoelectric effect – in the photoelectric effect, electrons are emitted from solids, liquids or gases when they absorb energy from light

Photoelectron – an electron ejected from a material by the photoelectric effect

Photon – a quantum of electromagnetic radiation

Planck's constant – the constant of proportionality between the energy of a photon and its frequency, with a value of 6.63×10^{-34} J s

Point source – where the light from a point spreads out evenly in all directions

Threshold frequency – the minimum frequency of a photon which causes electron emission

Work function – the minimum energy required to eject an electron from its parent atom

Zero potential energy – an electron has zero potential energy in a nucleus' field when it has escaped from the field; this happens when ionisation takes place

Questions

Use these data in the following three questions:

The charge on an electron = 1.6×10^{-19} C

The mass of an electron = 9.11×10^{-31} kg

Plank's constant = 6.63×10^{-34} J s

1 Light is shone on to a metal plate. No photoemission takes place.
   a) The irradiance of the light is increased. What happens? Justify your answer.
   b) The wavelength of the light is now increased. What happens? Justify your answer.
2 A metal plate has a work function of 2.65×10^{-18} J.
   a) Calculate the threshold frequency for this metal plate.
   b) Photons of frequency 5.6×10^{15} Hz are shone onto the metal plate. What is the maximum kinetic energy of the photoelectrons emitted?
3 A metal plate has a threshold frequency of 1.2×10^{15} Hz. Light with a wavelength of 167 nm is shone onto the plate. Calculate:
   a) the frequency of the photons of light
   b) the energy of the photons
   c) the maximum kinetic energy of the emitted photoelectrons
   d) the maximum speed of the emitted photoelectrons.

4 State what is meant by irradiance and give the SI unit for irradiance.

5 A solar panel of area $4.0\,m^2$ collects $1200\,J$ of energy in 1 minute. Calculate the irradiance on the solar panel.

6 A point source of light is set up. The irradiance of this source is $64\,W\,m^{-2}$ at a distance of $1.25\,m$ from the source. What will be the irradiance $2.50\,m$ from the source?

7 Draw the Bohr model for a hydrogen atom. Your diagram should include labels for the ground state and the first three excited states.

8 Figure 12.9 represents the energy levels in an element.

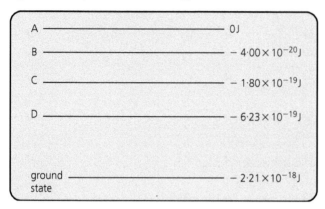

A	————————————	0 J
B	————————————	-4.00×10^{-20} J
C	————————————	-1.80×10^{-19} J
D	————————————	-6.23×10^{-19} J
ground state	————————————	-2.21×10^{-18} J

Figure 12.9 Energy levels in an element.

   a) (i) How many energy transitions are possible between energy level B and the ground state?
   (ii) How many spectral lines will the transitions in (i) produce?
   b) (i) Calculate the energy released when an electron moves from level C to level D.
   (ii) Calculate the frequency of the photon produced when an electron moves from level C to level D.
   c) Calculate the wavelength of the photon with the longest wavelength that this element will produce.

9 Some of the energy levels of Hydrogen are given as shown below.

E_3 ———————— -1.360×10^{-19} J

E_2 ———————— -2.416×10^{-19} J

E_1 ———————— -5.424×10^{-19} J

E_0 ———————— -21.760×10^{-19} J

Figure 12.10 Energy levels of Hydrogen

   a) Calculate the wavelength of the energy emitted when a photon transitions from E_2 to E_0.
   b) The absorption of this wavelength produces a faint dark line in the continuous spectrum from the Sun. In which colour of the spectrum is this dark line observed?

10 UV radiation from a lamp is shone upon a strip of metal. The UV photons have an energy of 5.95×10^{-19} J. The work function of the metal is 2.94×10^{-19} J.
   a) Calculate:
   (i) the maximum kinetic energy of an electron released from this metal by this radiation
   (ii) the maximum speed of an emitted electron.
   b) An additional identical lamp is now shone on the strip. State the effect this has on the maximum speed of the electrons emitted.

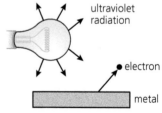

Figure 12.11 UV radiation from a lamp shone on metal

Chapter 13

Refraction of light

Refractive index

When an electromagnetic wave travels from one medium to another, the velocity of that wave can change. In general, when a wave goes from a less dense to a more dense medium, its velocity decreases. In dealing with light we are referring to the **optical density** of the medium which is not the same as its physical density. The optical density is a measure of the way the light interacts with the material as it passes through it. One indicator of the optical density of a material is its **refractive index**.

It had been noticed for many years that light changed direction when travelling from air to water, air to glass, etc. The light could appear to be distorted by travelling through different media.

The light turns *towards* the **normal** when going from air to glass and *away* from the normal when going from glass to air. This is due to light slowing down when in glass and speeding up when in air.

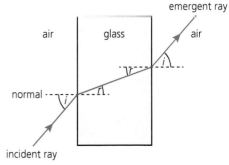

Figure 13.1 Refraction of light in a glass block

The refractive index of a material is the ratio of the sine of the angle of incidence to the sine of the angle of refraction:

$$n_{material} = \frac{\sin i}{\sin r}$$

We take the refractive index of air to be the same as that of a vacuum and this is equal to 1.

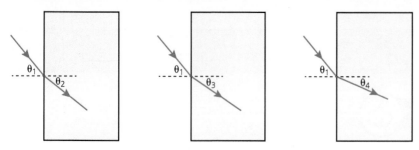

Figure 13.2 Different paths of light

The different paths of light in the blocks in Figure 13.2 are due to the blocks being made of different materials with differing refractive indices. The material with the largest refractive index causes the light to change direction most.

Example

Light is incident on a block as shown in Figure 13.3. From the information given in the diagram, calculate the angle of refraction, r.

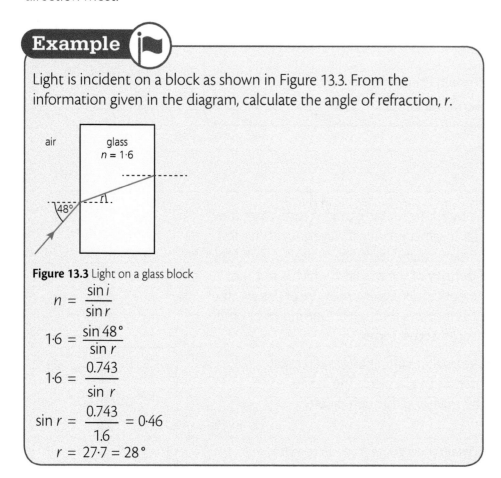

Figure 13.3 Light on a glass block

$$n = \frac{\sin i}{\sin r}$$

$$1{\cdot}6 = \frac{\sin 48°}{\sin r}$$

$$1{\cdot}6 = \frac{0.743}{\sin r}$$

$$\sin r = \frac{0.743}{1.6} = 0{\cdot}46$$

$$r = 27{\cdot}7 = 28°$$

When light exits glass and moves to a less dense material, we have to adapt our equation slightly.

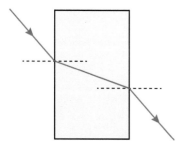

Figure 13.4 Light exiting a glass block

The ratio of the sines of the angles is also the ratio of the refractive indices, but as we take the refractive index of air to be 1, we ignore it in an algebraic sense.

$n = \dfrac{\sin i}{\sin r}$ becomes $\dfrac{n_2}{n_1} = \dfrac{\sin \theta_1}{\sin \theta_2}$

Rearranging gives:

$$n_2 \sin \theta_2 = n_1 \sin \theta_1$$

where the suffix 1 refers to the angle and refractive index in air (or medium 1) and the suffix 2 refers to the angle and refractive index in glass (or medium 2).

This relationship will allow us to calculate the change in direction for a ray of light going from oil to water to air to glass, for example. As long as we are consistent in our use of angles and refractive indices, the calculations are relatively straightforward.

Examples

Example of light leaving a glass block

Light strikes the internal surface of the glass block at an angle of $38\,^{\circ}$ to the normal. The refractive index of the glass, n, is 1.45. Calculate the angle at which the light exits the glass.

Solution

$$n_1 \sin \theta_1 = n_2 \sin \theta_2$$
$$1.45 \times \sin 38\,^{\circ} = 1 \times \sin \theta_2$$
$$0.89 = \sin \theta_2$$
$$\theta_2 = 63\,^{\circ}$$

Example of a triangular prism

Light is incident on a triangular prism at an angle of $60\,^{\circ}$ as shown in Figure 13.5. The refractive index of the glass prism is 1.5.

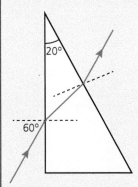

Figure 13.5 Light entering a triangular prism

Calculate the angle at which the light ray leaves the prism.

⇨
Solution

$n_1 \sin \theta_1 = n_2 \sin \theta_2$

$1 \times \sin 60° = 1.5 \times \sin \theta_2$

$0.866 = 1.5 \times \sin \theta_2$

$\frac{0.886}{1.5} = \sin \theta_2$

$\theta_2 = 35.3°$

Using the angles in the top triangle, the angle of incidence in glass

$= (180° - (54.7° + 20°)) = 105.3°$

$105.3° - 90° = 15.3°$

$n_1 \sin \theta_1 = n_2 \sin \theta_2$

$1.5 \sin 15.3° = 1 \sin \theta_2$

$0.396 = \sin \theta_2$

$\theta_2 = 23.3°$

Hints & tips

Care needs to be taken when working with angles in diagrams.

In equations, the angles used are always measured between the ray of light and the normal. Make sure that the angles on diagrams are measured from the normal – if not, you need to determine what the angle is before you use the equation.

Refractive index and frequency

The refractive index is not constant for all electromagnetic radiations; it varies with the frequency of the radiation (although a number of texts refer to the variation with **wavelength**). Different frequencies of light refract by different amounts and this leads to visible light 'splitting' into various colours. This effect is also referred to as **dispersion**.

If white light, for example, is refracted, the ray of white light disperses slightly. If we can make the light disperse enough, the visible spectrum of light can be seen as the colours begin to separate.

In general, the refractive index is greater for light of higher frequencies even within the same type of glass. For example:

$n_{\text{white light}} = 1.50$

$n_{\text{blue light}} = 1.53$

$n_{\text{red light}} = 1.48$

Example

A ray of red light and a ray of blue light are shone into a glass prism at an angle of incidence of 58°. The refractive indices of the light rays are as given above. Calculate the angle of refraction (θ_2) for each light ray.

Solution

Angle of incidence, $\theta_1 = 58°$

For red light: $n_1 \sin \theta_1 = n_2 \sin \theta_2$

$1 \times \sin 58° = 1.48 \times \sin \theta_2$

$0.57 = \sin \theta_2$

$\theta_2 = 35.0°$

For blue light: $n_1 \sin \theta_1 = n_2 \sin \theta_2$

$$1 \times \sin 58° = 1.53 \times \sin \theta_2$$
$$0.55 = \sin \theta_2$$
$$\theta_2 = 33.7°$$

The blue light is refracted more than the red. This separation of colours increases when the light undergoes further refraction at the other face of the prism and ultimately results in the production of a spectrum when white light is used. Other colours of light may be dispersed; it depends upon the frequencies of light that make up the initial colour of light.

The frequency of a ray of light is determined by the light source itself. It does not change as it goes from one medium to another. Consider a ray of frequency 5.0×10^{14} Hz entering a glass block.

That ray will exit the block at the same frequency; it cannot gain or lose 'waves' while passing through. As frequency remains constant, we can adapt the **wave equation**, $v = f\lambda$. Rearranging gives $f = v/\lambda$. As the frequency does not change when it goes from medium 1 to medium 2, the ratio v/λ must also be the same. Therefore:

$$\frac{v_1}{\lambda_1} = \frac{v_2}{\lambda_2}$$

This can be arranged to give

$$\frac{v_1}{v_2} = \frac{\lambda_1}{\lambda_2}$$

The change in angle is due to the change in refractive index which is in turn due to the change in velocity. This allows us to derive the following relationship:

$$\frac{\sin \theta_1}{\sin \theta_2} = \frac{v_1}{v_2} = \frac{\lambda_1}{\lambda_2} = n$$

This relationship, or part of it, can be used to calculate velocity, wavelength and direction of a ray of light due to refraction as it goes from one material to another.

Internal reflection and critical angle

In the earlier example of light travelling through the triangular prism, it can be seen that the angle the light makes with the normal when leaving the prism is greater than the angle it makes with the normal when inside the glass. This leads to a characteristic of light when travelling from glass to air. If light strikes the glass at a certain angle, the light will be refracted at 90°. This is known as the **critical angle**. Any angle of incidence greater than this angle will result in reflection without refraction. When all the light is reflected internally, this is known as **total internal reflection**. This critical angle is dependent upon the refractive index of the glass and can be calculated using $n_1 \sin \theta_1 = n_2 \sin \theta_2$.

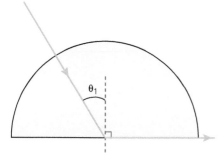

Figure 13.6 The angle of incidence is at the critical angle. The light is refracted at exactly 90°. The critical angle is defined as the angle of incidence that produces an angle of refraction of 90°.

If we take glass with a known refractive index, n, of 1·55, we can determine the angle at which the light refracts at 90°.

$$n_1 \sin \theta_1 = n_2 \sin \theta_2$$
$$1{\cdot}55 \times \sin \theta_1 = 1 \times \sin 90°$$
$$1{\cdot}55 \times \sin \theta_1 = 1 \times 1$$
$$\sin \theta_1 = \frac{1}{1{\cdot}55} = 0{\cdot}65$$
$$\theta = 40{\cdot}2°$$

This is the critical angle. It can also be calculated using

$$\theta_c = \sin^{-1}\left(\frac{1}{n}\right).$$

It is this property of refractive materials that allows data to be transmitted as pulses of light along fibre optic cables. If we ensure that the cables are relatively fine or narrow, the angles at which the light strikes the boundary will always be greater than the critical angle and the light will reflect internally along great distances.

Example

Experiment to measure refractive index

You need to be able to describe an experiment to measure the refractive index of a material.

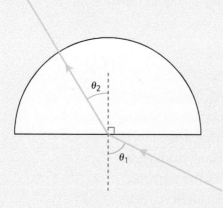

Figure 13.7 Experiment to measure the refractive index

The apparatus shown is set up with a normal drawn at the midpoint on the straight edge. A ray of light enters the block at an angle θ_1 to the normal. The ray of light that emerges from the block is traced so that angle θ_2 can be measured.

The experiment is repeated several times for different values of θ_1. The sine of both θ_1 and θ_2 are calculated and then plotted on a graph.

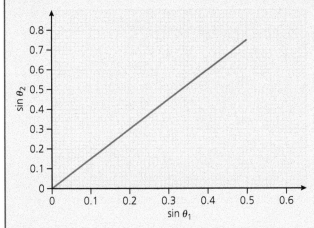

Figure 13.8 Graph measuring refractive index

The refractive index is the gradient of the best-fit line on the graph.

Key points

* When light travels from air into a **transparent** material, it slows down and its wavelength decreases. The frequency does not change.
* If the light enters the material at an angle to the normal, it changes direction towards the normal.
* The ratio of the sine of the angle in air to the sine of the angle in the material is constant. This constant is the refractive index of the material.
* The refractive index of a material is always greater than 1.
* When light travels from an optically more dense material into an optically less dense material (for example, when light travels from glass into air), it changes direction away from the normal. However, it can only do this as long as the angle of refraction in air is less than 90°. The angle of incidence in the material that produces an angle of refraction of 90° in air is called the critical angle.
* If the angle in the optically dense material is greater than the critical angle, the light is totally internally reflected.
* The refractive index of a material is equal to the ratio of the speed in air to the speed in the material. It is also equal to the ratio of the wavelength in air to the wavelength in the material.
* For a given material, the refractive index of the material increases as the frequency of the light increases.

Key words

Critical angle – the angle of incidence in the material that causes a ray of light to refract into air at 90°

Dispersion – visible light 'splitting' into various colours

Normal – a line drawn at right angles to a surface

Optical density – a measure of the way light interacts with a material as it passes through it

Refract – when a wave changes speed, wavelength and direction as it moves from one material to another

Refractive index – the ratio of the speed in air to the speed in the material

Total internal reflection – when all light is reflected internally within a transparent material

Transparent – allows light to pass

Wavelength – the minimum distance for a wave to repeat itself

Questions ?

1 Calculate the critical angle for materials with the refractive indices:
   a) $n = 1 \cdot 42$
   b) $n = 1 \cdot 67$
   c) $n = 2 \cdot 6$

2 A ray of light enters a block of transparent material as shown.

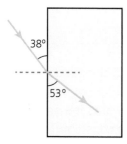

Figure 13.9 Ray of light in a glass block

   Determine the refractive index of the material.

3 When light passes from glass to air, state whether each of the quantities listed increases, decreases or stays the same.
   a) Angle relative to the normal
   b) Speed
   c) Wavelength
   d) Frequency

4 Explain why white light produces a spectrum of colours when it passes through a prism.

5 Ice has an absolute refractive index of 1·31 for yellow light. Calculate:
   a) the velocity of yellow light in ice
   b) the critical angle of yellow light in ice.

6 A ray of red light, wavelength 640 nm in air, enters a block of transparent material as shown.

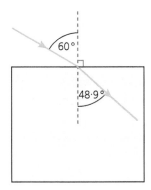

Figure 13.10 Ray of light through glass

Calculate:

a) the speed of light in the block

b) the wavelength of the red light in the block.

7 A ray of light has a wavelength of 4.75×10^{-7} Hz. This light is directed into a block of glass as shown. Calculate the frequency of the light in the block.

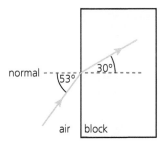

Figure 13.11 Path of light through glass

*8 Light travels from air into glass. Which row in the table describes what happens to the speed, frequency and wavelength of the light?

	Speed	Frequency	Wavelength
A	Increases	Decreases	Stays constant
B	Decreases	Stays constant	Decreases
C	Stays constant	Decreases	Decreases
D	Increases	Stays constant	Increases
E	Decreases	Decreases	Stays constant

Table 13.1 Table of speed, frequency and wavelength

*9 A student places a glass paperweight containing air bubbles on a sheet of white paper. The student notices that when white light passes through the paperweight, a pattern of spectra is produced. The student decides to study this effect in more detail by carrying out an experiment in the laboratory. A ray of green light follows the path shown as it enters an air bubble inside the glass.

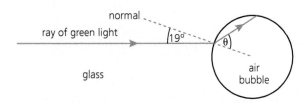

Figure 13.12 Ray of light through an air bubble inside glass

103

The refractive index of the glass for this light is 1·49.
- **a)** Calculate the angle of refraction, θ, inside the air bubble.
- **b)** Calculate the maximum angle of incidence at which a ray of green light can enter the air bubble.
- **c)** The student now replaces the ray of green light with a ray of white light. Explain why a spectrum is produced.

*10 A technician investigates the path of laser light as it passes through a glass tank filled with water. The light enters the glass tank along the normal at C, then reflects off a mirror submerged in the water.

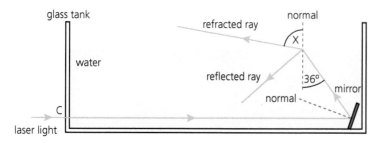

Figure 13.13 Path of laser light through a tank of water

The refractive index of water for this laser light is 1·33.

- **a)** Calculate angle X.
- **b)** The mirror is now adjusted until the light follows the paths shown.

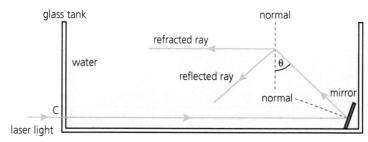

Figure 13.14 Path of laser light through a tank of water

  - **(i)** State why the value of θ is equal to the critical angle for this laser light in water.
  - **(ii)** Calculate angle θ.
- **c)** The water is now replaced with a liquid which has a greater refractive index. The mirror is kept at the same angle as in part **b)** and the incident ray again enters the tank along the normal at C. Draw a sketch which shows the path of the light ray after it has reflected off the mirror. Your sketch should only show what happens at the surface of the liquid.

Section 3 Electricity

Chapter 14
Electrons and electricity

What you should know

★ Alternating current is current which changes direction and instantaneous value with time
★ Calculations involving peak and r.m.s. values
★ Determination of frequency from graphical data
★ Use relationships involving potential difference, current, resistance and power to analyse circuits. Calculations may involve several steps
★ Calculations involving potential divider circuits
★ Electromotive force, internal resistance and terminal potential difference
★ Ideal supplies, short circuits and open circuits
★ Determining internal resistance and electromotive force using graphical analysis
★ Capacitors and the relationship between capacitance, charge and potential difference
★ The total energy stored in a charged capacitor is the area under the charge–potential difference graph
★ Using the relationships between energy, charge, capacitance and potential difference
★ Variation of current and potential difference with time for both charging and discharging
★ The effect of resistance and capacitance on charging and discharging curves

Current

Current is a flow of electrons caused by a potential difference (voltage) across a conductor. There are two types of current: **direct current** (DC) and **alternating current** (AC). In direct current, the electrons flow in one direction through a conductor. In alternating current, the electrons change direction back and forth in the conductor. Direct currents are produced by chemical cells and batteries. Alternating current is produced in UK power stations for mains electricity.

If you draw a graph of steady DC against time you would get the shape of graph shown in Figure 14.1.

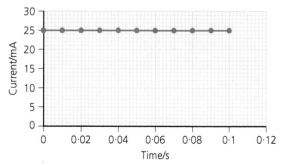

Figure 14.1 A graph of steady DC against time

If you draw a graph of AC against time you would get the shape of graph shown in Figure 14.2.

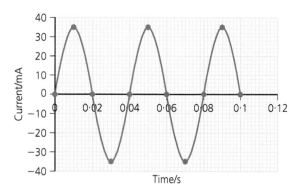

Figure 14.2 A graph of AC against time

It is important to notice that the value of an alternating current is changing with time. Because current and potential difference are proportional (**Ohm's Law**), the value of the potential difference must be changing with time as well.

Peak and root mean square (r.m.s.)

We can see from Figure 14.2 that there is a maximum value for current. This is called the **peak current**. It does not matter whether this value is positive or negative; it is still called the peak value (see Figure 14.3).

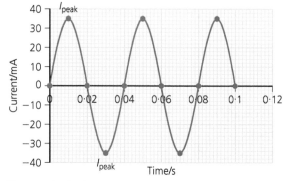

Figure 14.3 Peak current, I_{peak}

It is possible to draw the same graph for potential difference, as shown in Figure 14.4. The maximum potential difference is normally referred to as the **peak** voltage.

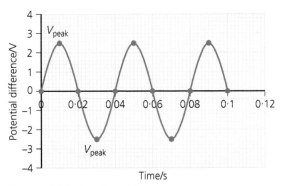

Figure 14.4 Peak voltage, V_{peak}

As AC is varying all the time, the current and potential difference values are continually changing. The value we give for alternating current and voltage is the equivalent value of a DC supply. For example, a bulb placed in a circuit with a 6·0 V DC supply will shine with the same brightness if it is placed in an identical circuit with a 6·0 V AC supply. The AC equivalent of a DC supply is known as the **r.m.s. value**. It stands for the root mean square and is the equivalent value of DC.

There is a mathematical relationship between the peak values and the r.m.s values for both current and potential difference:

$$I_{peak} = I_{r.m.s.} \times \sqrt{2}$$

and

$$V_{peak} = V_{r.m.s.} \times \sqrt{2}$$

Measuring frequency and peak voltage

It is possible to measure the frequency and peak voltage of an AC supply using an **oscilloscope**. You need to look at both the trace on the screen and the controls on the oscilloscope.

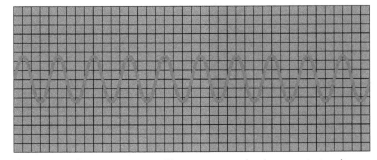

Figure 14.5 The trace on an oscilloscope screen (each square is 1 cm)

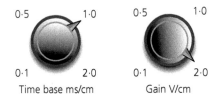

Time base ms/cm   Gain V/cm

Figure 14.6 The settings on an oscilloscope

The **time base** setting tells you the number of milliseconds it takes the trace to cross each box from left to right. The **voltage gain** tells you the number of volts each box represents in the vertical direction.

When we look at the trace in Figure 14.5 we can see that it takes four boxes across to make a complete cycle. This is equivalent to $4.0\,ms$ and is called the **period**, T, of the AC. We can use this to calculate the frequency of the AC using the following calculation:

$$f = \frac{1}{T}$$

$$f = \frac{1}{4 \times 10^{-3}}$$

$$f = 250\,Hz$$

This means that this AC supply has a frequency of $250\,Hz$.

We can also determine the peak voltage from the trace. The maximum height of the trace above the middle is 2·5 boxes. The voltage gain indicates that each box is worth 2·0 volts. The peak voltage is therefore $2.5 \times 2.0 = 5.0\,V$.

Now that we know the peak voltage we can calculate the r.m.s. voltage:

$$V_{r.m.s.} = V_{peak} \div \sqrt{2}$$

$$V_{r.m.s.} = 5.0 \div \sqrt{2}$$

$$V_{r.m.s.} = 3.55\,V$$

This AC supply can be described as a 3·55 V, 250 Hz supply.

Hints & tips ⭐

Care is needed when answering questions comparing AC and DC. The r.m.s. voltage is the equivalent of the DC voltage. You may find a multiple-choice question based on this. If you are asked, for example, what the DC equivalent of $2.0\,V_{r.m.s.}$ is the answer is simply $2.0\,V$.
The power of an appliance is calculated using the r.m.s. values of current and voltage.

Current, potential difference, power and resistance

In National 5 you studied simple series and parallel circuits and performed calculations in order to determine the current and voltage in and across various components in those circuits. At Higher you are expected to be able to apply the same principles but to more complex circuits.

Basic principles

- Voltages across components in a series circuit combine to give the supply voltage.
- The voltage across the branches of a parallel circuit is the same.
- The current in a series circuit is the same at all points.
- The currents in parallel circuit branches combine to give the current from the supply.

Relationships used in this topic include:

$$V = IR$$

$$P = IV = I^2R = \frac{V^2}{R}$$

$$R_T = R_1 + R_2 + \ldots$$

$$\frac{1}{R_T} = \frac{1}{R_1} + \frac{1}{R_2} + \ldots$$

$$\frac{V_1}{V_2} = \frac{R_1}{R_2}$$

$$V_1 = \left(\frac{R_1}{R_1 + R_2}\right) \times V_S$$

Examples

1 A circuit is set up as shown in Figure 14.7. Calculate:
   a) the voltage across the 180 Ω resistor
   b) the power dissipated in the 300 Ω resistor.

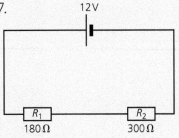

Figure 14.7 Series circuit with resistors

Solution

a) To calculate the voltage across one resistor, we need to calculate the current in that resistor and then use Ohm's Law:

$R_T = R_1 + R_2 = 180\,\Omega + 300\,\Omega = 480\,\Omega$

$I = \dfrac{V}{R} = \dfrac{12}{480} = 0{\cdot}025\,A$

$V = I \times R = 0{\cdot}025 \times 180 = 4{\cdot}5\,V$

b) To calculate the power dissipated in the 300 Ω resistor, we could use a number of methods.

$P = I^2 \times R = 0{\cdot}025^2 \times 300 = 0{\cdot}1875 = 0{\cdot}19\,W$

We could also determine the voltage across the 300 Ω resistor by simply subtracting the voltage across the 180 Ω resistor from the supply voltage: $V_2 = 12 - 4{\cdot}5 = 7{\cdot}5\,V$. Then use $P = IV$.

This circuit is a type of **potential divider** circuit. The potential difference of the supply is divided between the resistors in proportion to their resistances, *not* their values.

2 Calculate the potential difference across each resistor in the circuits shown in Figure 14.8.

a)

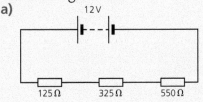

b)

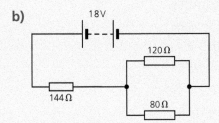

c)

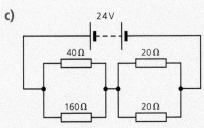

Figure 14.8 Series of circuits

⇨
Solution

a) $R_T = 1000\,\Omega$; $I = \dfrac{V}{R} = \dfrac{12}{1000} = 0.012\,A$

$V_1 = 0.012 \times 125 = 1.5\,V$

$V_2 = 0.012 \times 325 = 3.9\,V$

$V_3 = 0.012 \times 550 = 6.6\,V$

b) Total resistance of parallel section $= 48\,\Omega$

Total circuit resistance $= 192\,\Omega$

Total circuit current $= \dfrac{V}{R} = \dfrac{18}{192} = 0.09\,A$

P.d. across $144\,\Omega$ resistor $= I \times R = 0.09 \times 144 = 13.5\,V$

P.d. across other two resistors $= 18 - 13.5 = 4.5\,V$

c) Total resistance of left-hand pair $= 32\,\Omega$

Total resistance of right-hand pair $= 10\,\Omega$

Total circuit resistance $= 42\,\Omega$

Total circuit current $= \dfrac{V}{R} = \dfrac{24}{42} = 0.571\,A$

Current in each $20\,\Omega$ resistor $= 0.571 \div 2 = 0.286\,A$

P.d. across each $20\,\Omega$ resistor $= I \times R = 0.286 \times 20 = 5.7\,V$

P.d. across other two resistors $= 24 - 5.7 = 18.3\,V$

Applications of series and parallel circuits

Potential divider circuits

These circuits are used to provide or supply a specific voltage to a component. They rely upon the principle that the voltage is shared between components depending upon the ratio of their resistances. A series circuit with a $40\,\Omega$ and $80\,\Omega$ resistor will 'divide' the voltage from the supply in the ratio of $1:2$. Similarly a series circuit with a $1.5\,k\Omega$ and a $6\,k\Omega$ resistor will 'divide' the supply voltage in the ratio of $1:4$.

The voltage across and the current through each resistor is calculated in the same way as the previous examples.

Potential dividers often have a variable resistor of some sort as one of the resistors. This variable resistor could then change its resistance depending upon external conditions such as light or temperature. This would then alter the voltage across the components in the circuit which could trigger an alarm or indicate that something has occurred.

Electrical sources and internal resistance

Cells

An electrical cell produces a voltage due to a chemical reaction. This chemical reaction takes place between the metals and the acid or alkali contained inside the cell. During the chemical reaction, chemical potential energy is converted into electrical energy. Each coulomb of charge that passes through the cell is given a number of joules of energy by the reaction in the cell. This is called the **EMF** (or **electromotive force**), E, of the cell. As this is the number of joules per coulomb, the unit for EMF is the volt.

Until now we have regarded cells and other sources as a way of supplying a voltage to a circuit rather than as components in the circuit. However, cells are part of the circuit and, as we have seen, are made up of metal and other chemicals. These materials all have electrical resistance so a cell will have its own resistance. We call this the **internal resistance**, r, of the cell. We can think of a cell as being an EMF with an internal resistance in series.

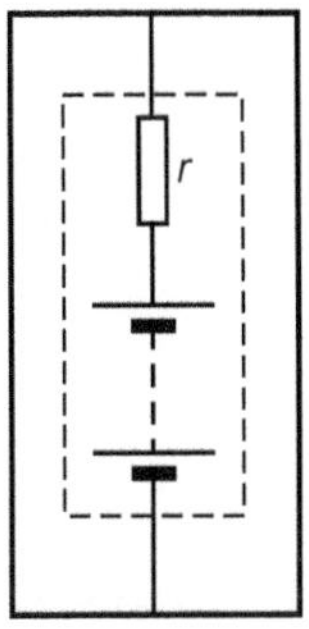

If we assume that a cell has no internal resistance we describe it as an ideal cell, or **ideal supply**.

Figure 14.9 A cell has an internal resistance, measured in ohms.

Cells in circuits

When we connect a cell in a circuit, a current flows. It is important to realise that this current also flows through the cell. The cell has an internal resistance and when current flows through it, a potential difference is developed across the internal resistance. This potential difference is lost inside the cell and is described as the **lost volts**, V_{lost}.

If the potential difference at the terminals of the cell is measured, it is found to be less than the EMF of the cell due to the lost volts. This is called the **terminal potential difference**, V_{tpd}.

Due to the principal of conservation of energy, the EMF of the cell must be equal to the terminal potential difference plus the lost volts:

$$E = V_{tpd} + V_{lost}$$

The number of lost volts in the cell depends on the size of the current:

$$V_{lost} = Ir$$

The larger the current, the higher the value for the lost volts. This means that the higher the current, the lower the terminal potential difference from the cell. This is why when we try to draw a large current from a cell, the voltage delivered drops.

Cells that have to deliver a high current need to have a very small internal resistance. For example, a car battery has a very low internal resistance so it can deliver a large enough current to start the starter motor.

Equations for EMF and lost volts

We already know that:

$$E = V_{tpd} + V_{lost} \text{ and } V_{lost} = Ir$$

We can combine these two equations to give:

$$E = V_{tpd} + Ir$$

We also know from Ohm's Law that $V = IR$, so:

$$V_{tpd} = IR$$

where R is the total circuit resistance and I is the circuit current.

We can now write some further equations:

$$E = IR + V_{lost}$$

$$E = IR + Ir$$

$$E = I(R + r)$$

All of these equations can be useful when carrying out calculations.

Measuring EMF and lost volts for a cell

A cell is connected in a circuit as shown in Figure 14.10.

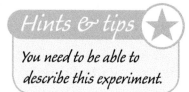

Hints & tips

You need to be able to describe this experiment.

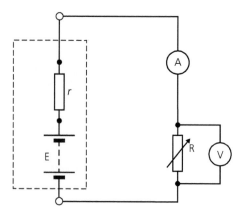

Figure 14.10 Series circuit diagram

The current in the circuit is altered using the variable resistor. The voltmeter measures the V_{tpd} of the battery. The results shown in Table 14.1 are obtained.

These results are used to produce a graph, as given in Figure 14.11.

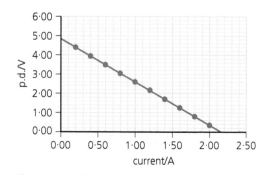

Figure 14.11 Experimental results for the circuit shown in Figure 14.10

I/A	V/V
0·20	4·40
0·40	3·95
0·60	3·50
0·80	3·05
1·00	2·60
1·20	2·15
1·40	1·70
1·60	1·25
1·80	0·80
2·00	0·35

Table 14.1

We can see from the graph that the terminal potential difference decreases as the current increases. There is a maximum terminal potential difference. This occurs when the current is zero. When the current is zero there are no lost volts: $V_{lost} = Ir$, so if $I = 0$ then $Ir = 0$. If there are no lost volts then the terminal potential difference is equal to the EMF of the battery.

To find the EMF of the battery we find the intercept of the line with the terminal potential difference axis. In this case the value is 4·85 V.

When there is no current flowing we describe this as an **open circuit**. The open circuit potential of a battery or a cell is the same as its EMF. We know that:

$$E = V_{tpd} + Ir$$

This can be rearranged to give:

$$r = \frac{E - V_{tpd}}{I}$$

This allows us to calculate the internal resistance of the battery from the graph in Figure 14.11:

$$r = \frac{4·85 - 0·35}{2·0} = 2·25\,\Omega$$

For this battery the EMF of the cell is 4·85 V and the internal resistance is 2·25 Ω.

We can also see from our graph that there is a maximum current the battery can deliver. This takes place when the external resistance of the circuit is zero and is called the **short circuit** current.

If we know the EMF of a cell and its internal resistance we can calculate the short circuit current. Thus:

$$E = I(R + r)$$

but R is zero so:

$$E = I_{short}r$$

For the battery in the experiment above:

$$I_{short} = 4·85 \div 2·25 = 2·16\,A$$

We can see that this is the intercept on the current (x) axis of our graph.

A graphical method can be helpful when trying to determine two unknown quantities.

Consider the equation:

$$E = I(R + r)$$

This leads to:

$$E = IR + Ir$$

113

Rearranging gives:

$$IR + Ir = E$$
$$IR = -rI + E$$

However, IR is the voltage from the battery, V, therefore:

$$V = -rI + E$$

If we draw a graph of V verses I, we get a straight line graph with a negative slope. We know that the equation for a straight line is $y = mx + c$, which is in the same form as our rearranged equation, $V = -rI + E$.

Comparing V with y and I with x gives the slope of the line as $-r$ and the intercept as E. This is a graphical solution to a problem measuring E and r.

Capacitors

Capacitance

Capacitors are common electronic devices used in many familiar circuits. They can be used to store charge in a circuit. They can also be used to store energy. If a capacitor is placed in a DC circuit, current will at first flow. Current is a flow of charge and this charge is stored on the capacitor.

If we measure the charge stored on the capacitor at different potential differences (voltages), we typically obtain results similar to those shown in Table 14.2 (the actual values will depend on the capacitor that is used).

Potential difference across capacitor/V	Charge on the capacitor/μC
1·00	5·00
2·00	10·00
3·00	15·00
4·00	20·00
5·00	25·00

Table 14.2

These results can be used to produce the graph in Figure 14.12.

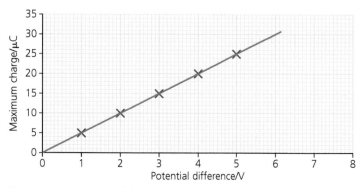

Figure 14.12 A graph of charge against potential difference

It can be seen that the results produce a straight line through the origin. This means that the charge stored on the capacitor is proportional to the potential difference.

If the charge stored is divided by the potential difference, the result is a constant for this capacitor. This constant is the **capacitance** (C) of the capacitor:

$$C = \frac{Q}{V}$$

where

C is the capacitance in farads (F)

Q is the charge in coulombs (C)

V is the potential difference in volts (V).

Capacitors typically have values measured in µF and nF.

Energy stored in a capacitor

The energy stored in a capacitor can be found from the area under the charge–potential difference graph shown in Figure 14.14. We can see that this graph has a triangular shape so the energy stored in the capacitor is given by the equation:

$$E = \frac{1}{2}QV$$

This equation can be combined with the capacitance equation to produce two more equations that allow the energy stored in a capacitor to be calculated:

$$E = \frac{1}{2}CV^2 \text{ and } E = \frac{1}{2}\frac{Q^2}{C}$$

> **Hints & tips** ⭐
>
> *Be careful when using this equation. It should only be used for energy stored in capacitors. If you are calculating the work done in moving a charge in an electric field or between charged plates you should use E = QV.*

Graphs for charging and discharging capacitors

When a capacitor is charged, a graph of current against time is obtained (Figure 14.13).

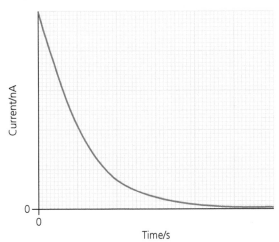

Figure 14.13 The current flowing in the circuit drops as the capacitor charges.

As the capacitor is initially uncharged, the electrons flow freely from the electrical supply to the capacitor resulting in a large current. As the electrons build up on one plate of the capacitor they exert a repulsive force on electrons coming onto the capacitor, thus reducing the rate of flow of electrons, so the current decreases. Eventually the current drops to zero and the capacitor is fully charged for this electrical supply voltage.

This also helps to explain the shape of the graph of potential difference against time (Figure 14.14).

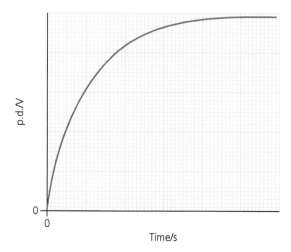

Figure 14.14 The potential difference across a capacitor as it charges

The current only flows when there is a potential difference between the supply and the capacitor. When the potential difference across the capacitor is the same as the supply, no more current flows. This means that the maximum voltage across the capacitor is the same as the voltage of the supply.

When a capacitor is discharged the electrons flow in the opposite direction to when the capacitor was charging. This means that the current will be in the opposite direction. Figure 14.15 shows how the current varies with time as a capacitor is discharged.

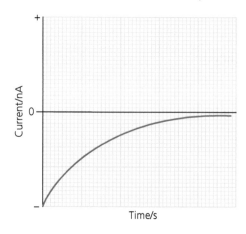

Figure 14.15 Graph of current versus time as a capacitor discharges

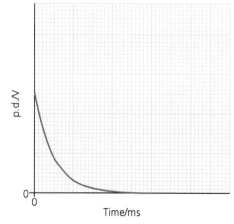

Figure 14.16 Graph of p.d. versus time as a capacitor discharges

Similarly, as a capacitor discharges, the potential difference across it drops until it becomes zero (Figure 14.16).

116

Example

Experiment to investigate the charging and discharging of a capacitor

You need to be able to describe an experiment to show how current and voltage in a circuit change as a capacitor charges and discharges.

The circuit shown below is used for this experiment.

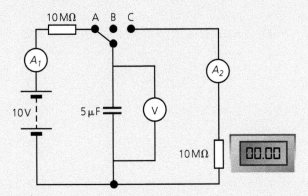

Figure 14.17 A simple camera flash circuit

The switch starts at position B and the capacitor is initially uncharged.

The switched is moved to position A and, at the same time, the stop clock is started. The capacitor begins to charge.

The reading on ammeter A_1 and the reading on the voltmeter are noted every five seconds.

Once the current falls to almost zero the switch is moved back to position B and the stop clock is switched off.

Graphs of current in the circuit against time, and voltage across the capacitor against time can be drawn for the charging of the capacitor.

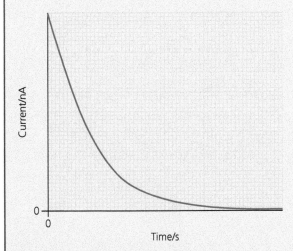

Figure 14.18 Graph of current vs time

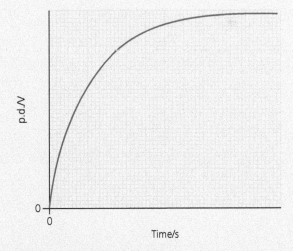

Figure 14.19 Graph of pd vs time

The stop clock is reset.

The switch is moved from position B to C and the stop clock is started again. The capacitor now discharges.

The reading on ammeter A_2 and the reading on the voltmeter are noted every five seconds.

Once the current falls to almost zero the switch is moved back to position B and the stop clock is switched off.

Graphs of current in the circuit against time, and voltage across the capacitor against time can be drawn for the discharging of the capacitor.

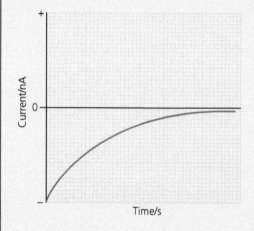

Figure 14.20 Graph of current vs time

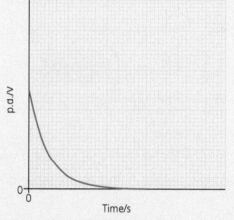

Figure 14.21 Graph of pd vs time

Changing resistance and capacitance in a circuit

One of the factors that determines how long it takes a capacitor to charge fully is the resistance of the circuit. A high-value resistor will limit the size of the current and, as current is the rate of flow of charge, it will take longer for the capacitor to become fully charged.

Similarly, if you increase the capacitance of the capacitor it will take longer to charge. This is because a capacitor with a higher capacitance can hold more coulombs of charge per volt, so it takes longer for the charge to fill the capacitor.

Consider the circuit shown in Figure 14.22. When the circuit is switched on, what will the current be?

At first there is no potential difference across the capacitor so all of the voltage will be across the resistor. We can calculate the current using $V = IR$:

$$I = \frac{V}{R} = \frac{6}{50 \times 10^3} = 0.12 \text{ mA}$$

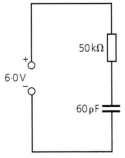

Figure 14.22 Circuit diagram

The current will now start to fall from this initial value.

Now consider this problem – what will be the charge stored on the capacitor at the instant when the current is 0·02 mA?

First we need to determine the potential difference across the resistor at this point:

$$V = IR = 0.02 \times 10^{-3} \times 50 \times 10^3 = 1.0 \text{ V}.$$

As the potential differences in a series circuit add up to the supply voltage, the potential difference across the capacitor is 5·0 V.

We can now calculate the charge stored on the capacitor at this instant:

$$Q = CV = 60 \times 10^{-12} \times 5 = 0.3 \text{ nC}$$

Key points !

* Alternating current (AC) is a current that changes direction.
* The value of AC changes with time.
* The value for potential difference (voltage) also changes with time.
* The maximum value of the current and the potential difference are called the peak values.
* The values of current and potential difference that are equivalent to DC current and potential difference are called the r.m.s. values.
* You should be able to use peak values to calculate r.m.s. values.
* You should be able to recongnise and draw AC traces on an oscilloscope.
* You should be able to calculate the frequency of an AC voltage from an oscilloscope trace.
* You should be able to calculate a peak potential difference from an AC oscilloscope trace.
* The electromotive force (EMF) of a source is the number of joules of electrical energy given to every coulomb of charge that passes through the source.
* The EMF is measured in volts.
* The EMF of a source is the maximum potential difference that a source can deliver. This is the potential difference when the current is zero and so is called the open circuit potential.
* When a current is drawn from the supply, the terminal potential difference is always less than the EMF of the source.
* The difference between the EMF and the terminal potential difference is called the lost volts.
* The lost volts are due to the internal resistance of the source.
* An electrical source can be thought of as a source of electrical potential (the EMF) in series with a resistor (the internal resistance).
* An ideal supply is an electrical supply that has no internal resistance.
* A short circuit is when two wires connect directly across a component. The current flows through the short rather than through the component, in effect removing it from the circuit. A short circuit has zero or very little resistance.
* An open circuit is when there is a gap in the circuit. This means that no current flows in the circuit. An open circuit has infinite or a very high resistance.
* You should be able to carry out an experiment to measure the EMF and internal resistance of a source.
* You should be able to calculate the EMF and internal resistance of a source from a voltage–current graph.
* Capacitors are devices that can store charge or energy.
* The capacitance of a capacitor is the charge stored on the capacitor divided by the potential difference across it.
* The capacitance of a capacitor can be found from the slope of the line on a charge against potential difference graph.
* The energy stored in a capacitor can be found from the area under a charge against potential difference graph.
* You should be able to use the equations $E = \frac{1}{2} QV = \frac{1}{2} CV^2 = \frac{1}{2} \frac{Q^2}{C}$ and $Q = CV$.
* You should be able to describe the current–time and voltage–time graphs when a capacitor charges and discharges.
* You should be able to describe the effect on the above graphs when either the resistance of the circuit or the capacitance of the capacitor is altered.

Key words

AC – alternating current
Capacitance – the ratio of charge to voltage for a capacitor; the number of coulombs of charge a capacitor can store per volt
Capacitor – an electronic device that can store charge and energy in a DC circuit
Current – a flow of charge (or electrons)
DC – direct current
Electromotive force (EMF) – the number of joules of electrical energy given to each coulomb of charge that passes through the source
Ideal supply – a supply that has no internal resistance
Internal resistance – the resistance of the material that makes up the cell
Lost volts – the potential difference across the internal resistance
Ohm's Law – the potential difference across a component is proportional to the current flowing through it, usually given by the equation $V = IR$
Open circuit – a circuit in which no current is drawn
Peak – the maximum current or voltage
Period – the time taken for one complete cycle or oscillation
Potential difference – the number of joules per coulomb; often called voltage between two points
r.m.s. – root mean square; the r.m.s. AC voltage is equivalent to the DC voltage
Short circuit – a path with zero electrical resistance
Terminal potential difference – the voltage measured across the terminals of a power supply
Voltage – another name for potential difference

Questions

1 In the UK the mains electrical supply is described as 230 V, 50 Hz.
   Calculate:
   a) the number of AC cycles per second
   b) the period of one cycle
   c) the peak voltage
   d) the r.m.s. current in a toaster heating element that has a resistance of 57·5 Ω
   e) the peak current in the toaster heating element.
2 Figure 14.23 shows the trace on an oscilloscope from an AC supply.

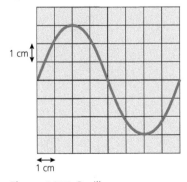

Figure 14.23 Oscilloscope trace

The voltage gain setting is 0·50 volts per centimetre. The time base setting is 0·10 milliseconds per centimetre. Calculate:

a) the frequency of the AC supply

b) the peak voltage of the supply

c) the r.m.s. voltage of the supply.

3 A kettle is connected to the mains electricity supply and is switched on. The current in the kettle is 12·0 A. Calculate:

a) the power of the element

b) the resistance of the element.

4 A box contains the following resistors: 12 Ω, 24 Ω and 48 Ω.

a) What arrangement of these resistors gives

  (i) the maximum resistance possible

  (ii) the minimum resistance possible?

b) How can the resistors be arranged to produce 56 Ω?

5 A circuit is set up as shown in Figure 14.24.

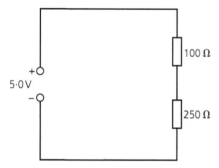

Figure 14.24 Circuit diagram

Calculate:

a) the potential difference across the 100 Ω resistor

b) the power dissipated by the 250 Ω resistor.

6 The circuit shown in Figure 14.25 is set up.

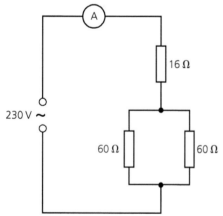

Figure 14.25 Circuit diagram

Calculate the potential difference across the 16 Ω resistor.

7 The circuit shown in Figure 14.26 is set up.

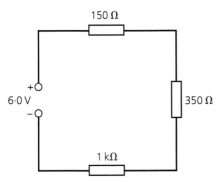

Figure 14.26 Circuit diagram

Calculate the total energy converted per second by the resistors.

8 What is meant by EMF?

9 What is the internal resistance of an ideal cell?

10 A battery has an EMF of 4·4 V and internal resistance 0·4 Ω. The battery is connected to a bulb with resistance 1·8 Ω.

   a) What is the current in the circuit?
   b) What is the terminal potential difference of the battery?
   c) What is the value of the lost volts for this battery?

11 A cell has an EMF of 1·5 V and internal resistance of 0·30 Ω. What is the short circuit current for this cell?

12 The following circuit is set up.

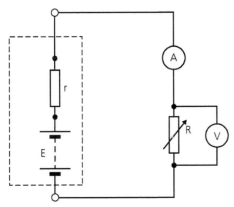

Figure 14.27 Circuit diagram

The resistance of the variable resistor is altered and the following results are obtained.

I/A	V/V
0·3	8·8
0·6	7·9
0·9	7·0
1·2	6·1
1·5	5·2
1·8	4·3
2·1	3·4
2·4	2·5
2·7	1·6
3·0	0·7

Table 14.3

a) Use these results to draw a graph of voltage against current for this battery.

b) Use the graph to determine the EMF of the battery.

c) Calculate the internal resistance of the battery.

13 A capacitor is fully charged using a 12-volt supply. The capacitor holds 42 nC of charge. Calculate:

a) the capacitance of the capacitor

b) the energy stored on the capacitor when it is fully charged.

14 Figure 14.28 shows a capacitor connected in a circuit.

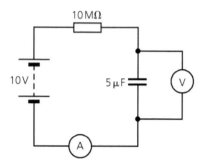

Figure 14.28 A capacitor connected in a circuit

The capacitor is initially uncharged. Sketch graphs to show how the readings on the voltmeter and ammeter vary with time as the capacitor charges. Indicate initial and final ammeter readings. Time values are not required.

15 The following circuit is set up.

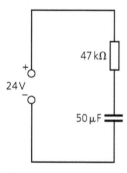

Figure 14.29 A capacitor connected in a circuit

The capacitor is initially uncharged. Calculate:

a) the initial charging current

b) the maximum charge stored on this capacitor for this supply

c) the energy stored on the capacitor when the potential difference across the resistor is 8·0 V.

*16 A microphone is connected to the input terminals of an oscilloscope.

A tuning fork is made to vibrate and held close to the microphone as shown.

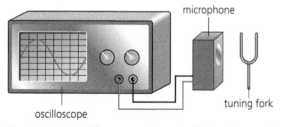

Figure 14.30 Oscilloscope with tuning fork and microphone

The following diagram shows the trace obtained and the settings on the oscilloscope.

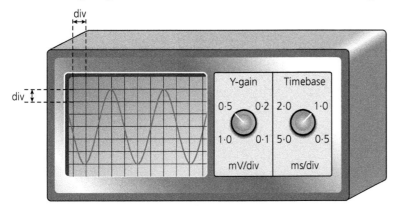

Figure 14.31 Oscilloscope trace and settings

Calculate:
   a) the peak voltage of the signal
   b) the frequency of the signal.

*17 A signal generator is connected to a lamp, a resistor and an ammeter in series. An oscilloscope is connected across the output terminals of the signal generator.

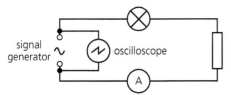

Figure 14.32 Oscilloscope in a circuit diagram

The oscilloscope control settings and the trace displayed on its screen are shown.

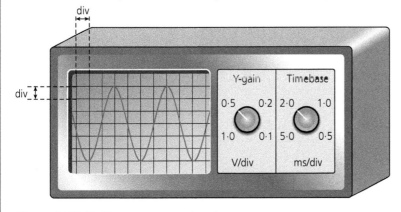

Figure 14.33 Oscilloscope trace and settings

   a) For this signal, calculate:
      (i) the peak voltage
      (ii) the frequency.
   b) The frequency is now doubled. The peak voltage of the signal is kept constant. State what happens to the reading on the ammeter.

*18 A thermocouple is a device that produces an EMF when heated.

a) A technician uses the circuit shown to investigate the operation of a thermocouple when heated in a flame.

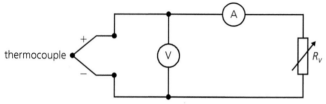

Figure 14.34 Thermocouple in a circuit diagram

Readings of current and potential difference (p.d.) are recorded for different settings of the variable resistor, R_v. The graph of p.d. against current is shown.

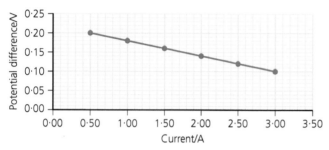

Figure 14.35 Graph of pd against current

Use information from the graph to find:

(i) the EMF produced by the thermocouple

(ii) the internal resistance of the thermocouple.

b) A different thermocouple is to be used as part of a safety device in a gas oven. The safety device turns off the gas supply to the oven if the flame goes out. The thermocouple is connected to a coil of resistance of $0.12\,\Omega$ which operates a magnetic gas valve.

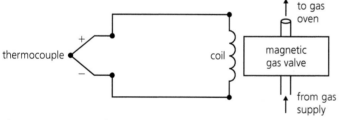

Figure 14.36 Circuit diagram of thermocouple and magnetic gas valve

When the current in the coil is less than 2·5 A, the gas valve is closed. The temperature of the flame in the gas oven is 800 °C. The manufacturer's data for this thermocouple are shown in the two graphs.

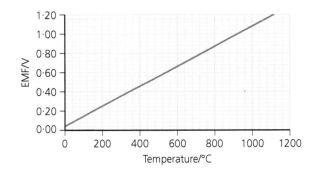

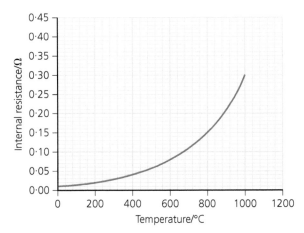

Figure 14.37 Two line graphs

Is this thermocouple suitable as a source of EMF for the gas valve to be open at a temperature of 800 °C? You must justify your answer.

*19 A student carries out an experiment using a circuit which includes a capacitor with a capacitance of 200 μF.

a) Explain what is meant by 'a capacitance of 200 μF'.

b) The capacitor is used in the circuit shown to measure the time taken for a ball to fall vertically between two strips of metal foil.

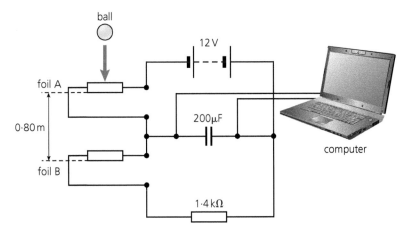

Figure 14.38 Capacitor in a circuit diagram

The ball is dropped from rest above foil A. It is travelling at $1.5\,\text{m}\,\text{s}^{-1}$ when it reaches foil A. It breaks foil A, then a short time later breaks foil B. These strips of foil are $0.80\,\text{m}$ apart. The computer displays a graph of potential difference across the capacitor against time as shown.

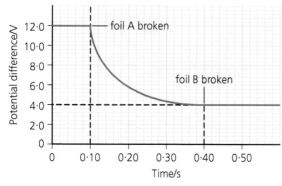

Figure 14.39 Line graph of pd against time

(i) Calculate the current in the $1.4\,\text{k}\Omega$ resistor at the moment foil A is broken.
(ii) Calculate the decrease in the energy stored in the capacitor during the time taken for the ball to fall from foil A to foil B.

c) The measurements from this experiment are now used to estimate the acceleration due to gravity.
   (i) What is the time taken for the ball to fall from foil A to foil B?
   (ii) Use the results of this experiment to calculate a value for the acceleration due to gravity.
   (iii) The distance between the two foils is now increased and the experiment repeated. Explain why this gives a more accurate result for the acceleration due to gravity.

Chapter 15
Electrons at work

What you should know

★ Solids can be categorised into conductors, semiconductors or insulators by their ability to conduct electricity

★ The electrons in atoms are contained in energy levels. When the atoms come together to form solids, the electrons then become contained in energy bands separated by gaps

★ In metals, the highest occupied band is not completely full and this allows the electrons to move and therefore conduct. This band is known as the conduction band. For a solid to be conductive, both free electrons and accessible empty states must be available

★ For a solid to be conductive, both three electrons plus accessible empty states must be available

★ For metals we have the situation where one or more bands are partially filled. Some metals have free electrons and partially filled valence bands, therefore they are highly conductive. Some metals have overlapping valence and conduction bands. Each band is partially filled and therefore they are conductive

★ In a semiconductor, the gap between the valence band and conduction band is smaller and at room temperature there is sufficient energy available to move some electrons from the valence band into the conduction band, allowing some conduction to take place. An increase in temperature increases the conductivity of a semiconductor

★ During manufacture, the conductivity of semiconductors can be controlled, resulting in two types: p-type and n-type

★ When p-type and n-type materials are joined, a layer is formed at the junction. The electrical properties of this layer are used in a number of devices

★ Solar cells are p–n junctions designed so that a potential difference is produced when photons enter the layer. This is the photovoltaic effect

★ LEDs are p–n junctions which emit photons when a current is passed through the junction

★ Electrons move from the conduction band of the n-type material towards the conduction band of the p-type material. These electrons can fall from the conduction band on either side of the junction to the valence band. They lose energy when they do this and photons are emitted

Conductors, semiconductors and insulators

Conductors and insulators

In electrical terms, solids can be categorised into three groups by their ability to allow electron movement.

- **Conductors** allow electrons to move relatively easily.
- **Insulators** do not allow electrons to move easily.
- **Semiconductors** can act as conductors or insulators under certain conditions.

To explain this more clearly we need to reconsider the structure of an atom. Electrons orbit the nucleus of an atom but in certain, defined levels. In a conductor, the outer level of the atom contains electrons but not enough to fill it. This allows a degree of movement for the electrons. In an insulator, the outer level is full and the electrons are fixed or tied in the one level.

When the atoms of a material are brought together, the outer energy levels interact with each other to form an energy band. In metals, one or more of these bands is partially filled. In some metals, for example, electrons can move from atom to atom within this outer band as it is not completely full. This allows electrical conduction and so this energy band is known as the **conduction band**. In order for conduction to take place, there must be electrons in this band. The band level below this conduction band is called the **valence band** and, because it is full, it does not allow transfer or movement of charge within it. At room temperature, however, the valence band actually overlaps slightly with the conduction band and this allows electrons to move out to the conduction band and thus assist conduction. The band gap is negligible in energy terms.

In other metals there are free electrons and partially filled valence bands. These metals are highly conductive.

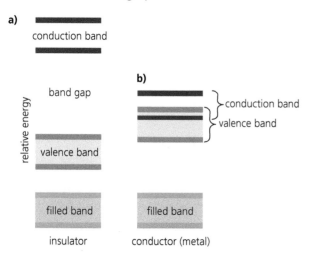

Figure 15.1 Energy bands in an insulator and in a conductor

In insulators, the outer level is full and there is no transfer or movement of electrons and no movement of charge. This band is known as the valence band. There is a conduction band at a level above this but it is empty with no free electrons. The gap between the valence and conduction band is large (see Figure 15.1) and so it would require a huge amount of energy to move an electron from the valence band to the conduction band. This band gap makes such electron movement difficult or unlikely.

If we provide enough energy to an insulator, in the form of a large voltage, for example, electrons can move from the valence band to the conduction band and it will conduct. This is very important in dealing with power supply lines where the voltage is measured in kV. At these voltages even very highly insulating materials can conduct and so the insulators that connect to our national grid have to be monitored. The voltage at which electrons in an insulator leave the valence band and jump to the conducting band is referred to as the **breakdown voltage**. A graphic example of this is lightning or high-voltage sparks in electrical transformers.

Semiconductors

Semiconductors are materials which are halfway between conductors and insulators. The band gap between valence and conduction in a semiconductor is not as large as that in an insulator, nor does it overlap as in a conductor. Semiconductors are designed so that small amounts of energy, due to room temperature or light, can give electrons enough energy to make the move and conduct. If the temperature of a semiconductor is increased it will conduct more easily.

Thermistors, for example, work because raising the temperature of a semiconductor increases the number of active charge carriers – it promotes them into the conduction band. The more charge carriers that are available, the more current a material can conduct. In certain materials like ferric oxide (Fe_2O_3) with titanium (Ti) doping, an n-type semiconductor is formed and the charge carriers are electrons. In materials such as nickel oxide (NiO) with lithium (Li) doping, a p-type semiconductor is created where 'holes' are the charge carriers. These types of semiconductor are discussed in greater depth in the next section.

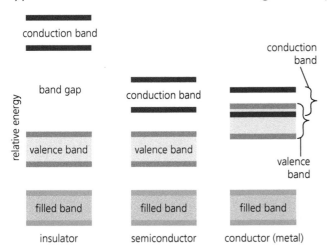

Figure 15.2 In a semiconductor, some electrons can occupy the conduction band at room temperature

p–n junctions

Most semiconductors are made from silicon, which has a valency of four (four electrons in the outer level). Silicon has a lattice structure. If we introduce an atom with a valency of five (five electrons in the outer level) into the silicon, we will have a very similar lattice structure but with an atom that has an additional electron. This electron can displace electrons from the silicon atoms and these electrons can displace other electrons in turn. This electron adds a negative charge and so this is referred to as an **n-type semiconductor**.

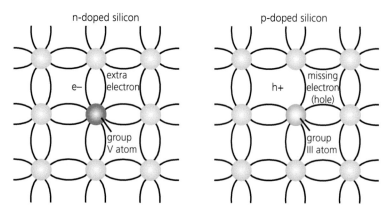

Figure 15.3 The crystalline structure of n- and p-type silicon

We can also insert an atom of valency three (three electrons in the outer level) which leaves a gap or hole where an electron would be. There is a missing electron in the lattice so this section of the lattice is more positive than it would otherwise be. This type of conductor is called a **p-type semiconductor**. Rather than the electron moving in one direction, the hole can appear to move by a series of electrons moving in the opposite direction.

The adding of impurities to pure silicon to produce n-type or p-type silicon is called **doping**. Doping alters the electrical properties of the silicon by increasing its conductivity (hence reducing its resistance).

The important part of doping in physics terms is what happens when we combine an n-type and a p-type semiconductor. When this is carried out, a layer forms at the junction and the properties of this layer can be adapted to our purposes.

The forward-bias voltage causes electrons in the conduction band of the n-type material to move towards the conduction band of the p-type material. Electrons can then drop from the conduction band into the valence band. As this happens, each electron that falls from the conduction band to the valence band causes an electron to be emitted.

Solar cells

When light is incident upon the depletion layer, it can transfer energy to an electron and it will move from the valence band to the conduction band. This movement of charge creates a potential difference and a small current will flow. If we arrange large areas of these so-called p–n junctions, the **solar cell** is referred to as a solar panel and a reasonably large current will flow (in only one direction, however).

a)

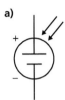

b)

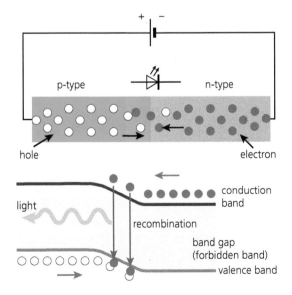

Figure 15.4 a) The circuit symbol for a solar cell; **b)** A solar panel on a house

Light-emitting diodes (LEDs)

LEDs operate in a way that is almost the reverse of the operation of a solar cell. In a solar cell, the photon of light moves the electron and generates a current. In an LED, when current is flowing across the junction, electrons pass into holes in the layer. Electrons in the conduction band of the n-type material are given energy by the forward-bias voltage from the power source. This causes them to move towards the conduction band of the p-type material. As they pass through the junction, some of them fall into the valence band. When they do this they lose energy. This energy is emitted as a photon.

Figure 15.5 A p–n junction

The amount of energy released determines the colour of light emitted by the LED. Red light is the lowest frequency and therefore requires the least amount of energy. This means that red LEDs, have a lower switch-on potential (voltage) than blue LEDs, as red photons have less energy than blue photons. In the past, it was difficult to make LEDs emit higher frequencies of light, but they are now commonplace.

a)

b)

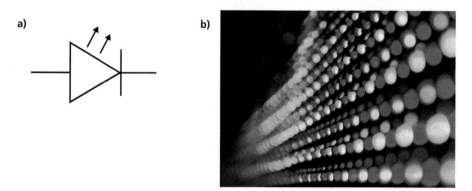

Figure 15.6 a) The circuit symbol for an LED; **b)** LEDs in use

Key points !

* Solids can be split into three groups according to their electrical properties: conductors, insulators and semiconductors.
* Conductors have a low resistance.
* Insulators have a high resistance.
* Pure semiconductors have a high resistance but this can be controlled by the introduction of impurities.
* When atoms form a solid, the arrangement of the atoms causes the electrons to be contained in energy bands. There are gaps between the energy bands.
* For a solid to conduct there must be free electrons and accessible empty states.
* In metals, one or more bands are only partially filled. In some metals the valence bands are only partially filled and so are highly conductive. Other metals have overlapping valence and conduction bands and therefore can conduct.
* In insulators, the highest occupied band is called the valence band and this band is completely full. The first unfilled band above the valence band is the conduction band. At room temperature, electrons do not gain enough energy to move from the valence band into the conduction band. This is why an insulator does not conduct.
* In a semiconductor, the energy gap between the valence band and the conduction band is much smaller. It therefore requires much less energy to move an electron from one band to the other. This means that some conduction can take place.
* When the temperature of a semiconductor increases, its conductivity increases (its resistance therefore decreases).
* p-type semiconductor material contains impurities that causes there to be a small excess of positive charge carriers in the material.
* n-type semiconductor material contains impurities that causes there to be a small excess of negative charge carriers in the material.
* When p-type and n-type semiconductor material is joined together, a layer is formed at the junction. The properties of this layer are used by various electronic devices such as LEDs and photodiodes.
* Solar cells use this layer to produce a voltage when photons enter the junction. This is known as the photovoltaic effect.
* LEDs emit photons of light when a current is passed through the junction.
* In metals, one or more bands is/are only partially filled. In some metals, the valence bands are only partially filled and so are highly conductive. Other metals have overlapping valence and conduction bands and, therefore, can conduct.

Key words

Conduction band – the band that allows conduction to take place; in conductors this band is partly filled

Conductor – a material that allows electricity to flow

Doping – the adding of impurities to pure silicon to produce n-type or p-type silicon

Insulator – a material that blocks the flow of electricity

LED – a light-emitting diode; a semiconductor device that gives out photons when current passes through it

n-type semiconductor – semiconductor material in which the majority of charge carriers are negative

p-type semiconductor – semiconductor material in which the majority of charge carriers are positive

Semiconductor – a material that will conduct under certain conditions; in semiconductors the gap between the valence band and the conduction band is much narrower than in insulators

Solar cell – a semiconductor device that produces a voltage when photons enter the semiconductor junction

Valence band – the highest occupied band in insulators; in insulators this band is full

Questions

1 Using band theory, draw a diagram to show the difference between conductors and insulators. Your diagram should show the conduction band, the valence band and the gap between for both types of material.

2 State what happens to the resistance of pure semiconductor material when an impurity is added.

3 What happens to the resistance of doped semiconductor material when its temperature increases?

4 Draw a circuit diagram to show how an LED should be connected to a supply and a resistor so that it will emit light. Explain the purpose of the resistor in the circuit.

Uncertainties

What you should know

★ SI units should be used with all the physical quantities
★ Prefixes should be used where appropriate. These include pico (p), nano (n), micro (μ), milli (m), kilo (k), mega (M), giga (G) and tera (T)
★ In carrying out calculations and using relationships to solve problems, it is important to give answers to an appropriate number of significant figures. This means that the final answer can have no more significant figures than the value with the least number of significant figures used in the calculation
★ Candidates should be familiar with the use of scientific notation and this may be used as appropriate when large and small numbers are used in calculations
★ All measurements of physical quantities are liable to uncertainty, which should be expressed in absolute or percentage form. Random uncertainties occur when an experiment is repeated and slight variations occur. Scale-reading uncertainty is a measure of how well an instrument scale can be read. Random uncertainties can be reduced by taking repeated measurements
★ Systematic uncertainties occur when readings taken are either all too small or all too large. They can arise due to measurement techniques or experimental design
★ The mean of a set of readings is the best estimate of a 'true' value of the quantity being measured. When systematic uncertainties are present, the mean value of measurements will be offset. When mean values are used, the approximate random uncertainty should be calculated
★ When an experiment is being undertaken and more than one physical quantity is measured, the quantity with the largest percentage uncertainty should be identified and this may often be used as a good estimate of the percentage uncertainty in the final numerical result of an experiment. The numerical result of an experiment should be expressed in the form
final value ± uncertainty

Units, prefixes and scientific notation

When you write down a measurement you should always include the unit. This is part of your answer.

Table 1 shows the common prefixes that are used and how to convert the quantity into the base unit using scientific notation.

Prefix	Factor
p (pico)	$\times 10^{-12}$
n (nano)	$\times 10^{-9}$
μ (micro)	$\times 10^{-6}$
m (milli)	$\times 10^{-3}$
k (kilo)	$\times 10^{3}$
M (mega)	$\times 10^{6}$
G (giga)	$\times 10^{9}$
T (tera)	$\times 10^{12}$

Table 1

Hints & tips ★

Remember that kilogram is the base unit and so does not have a factor applied when you are using it.

Significant figures

Be careful when you answer questions in the exam that you have used the same number of **significant figures** as in the question. In other words, it is important to round off your answers and not copy down exactly what it says on your calculator.

Hints & tips ★

When you are doing multiple-stage calculations, you should only round off at the end of the question in your final answer. Keep all the figures in your calculator as you go along, as this will increase the accuracy of your answer. It is fine to write down too many significant figures as part of your working as long as you round off in your final answer.

Example ⚑

A resistor has a current of 6·0 mA in it and a potential difference of 1·4 V across it. What is the resistance of the resistor?

Solution

$$R = \frac{V}{I} = \frac{1 \cdot 4}{6 \times 10^{-3}} = 233 \cdot 33333333333 \, \Omega$$

But, since there are only two significant figures in the numbers in the question, we should answer using two significant figures. Therefore, resistance, $R = 230 \, \Omega$.

Uncertainties

Every time a measurement is made it will have an uncertainty in it. Consider the circuit shown in Figure 1.

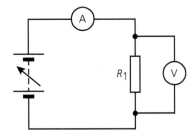

Figure 1

If you use a digital voltmeter to measure the potential difference across the resistor and obtain a reading of 6·6 volts, there is a **scale-reading uncertainty** of ±1 in the last digit. The reading on the meter is therefore 6·6 ± 0·1 V. When written like this the uncertainty is called the **absolute uncertainty**.

The **percentage uncertainty** in this reading is:

$$\frac{0.1}{6.6} \times 100 = 1.5\%$$

Suppose the reading on the digital ammeter is 0.21 A.

This reading should be recorded as 0.21 ± 0.01 A.

The percentage uncertainty in this reading is:

$$\frac{0.01}{0.21} \times 100 = 4.8\%$$

We can now compare the two readings. The percentage uncertainty in the potential difference is less than the percentage uncertainty in the current. We can use this percentage uncertainty when we work out the resistance of the resistor.

$$R = \frac{V}{I} = \frac{6.6}{0.21} = 31.42857143\,\Omega$$

But there are only two significant figures in the voltmeter reading, so there should only be two significant figures in our answer: $R = 31\,\Omega$.

We can now work out the uncertainty in the resistance.

The percentage uncertainty is the largest percentage uncertainty in the readings = 4.8%

$$R = 31\,\Omega \pm 4.8\%$$

The absolute uncertainty in R is:

$$31 \times 4.8\% = 1.488\,\Omega$$

But we need to be careful here. We have measured the resistance to the nearest whole number, so the absolute uncertainty must also be quoted to the nearest whole number.

$$R = 31 \pm 1\,\Omega$$

This is the final result of the experiment including the absolute uncertainty.

Random uncertainties come about when we make repeated measurements of the same thing. For example, if we use a stop clock to measure the time for a ball to run down a slope we might obtain the following results: 6.2 s, 6.4 s, 6.7 s, 6.5 s, 6.4 s, 6.3 s.

We can use these figures to calculate the average time:

$$\frac{total}{number} = \frac{6.2 + 6.4 + 6.7 + 6.5 + 6.4 + 6.3}{6} = 6.416666667\,s$$

But all of our readings are to two significant figures so the average time = 6.4 s.

The random uncertainty $= \frac{max\ reading - min\ reading}{number\ of\ readings} = \frac{6.7 - 6.2}{6} = \frac{0.5}{6} = 0.083\,s$.

All our times are quoted to one decimal place so our uncertainty should be the same = 0.1 s.

The average time is therefore 6.4 ± 0.1 s.

If we do more readings, we will be dividing by a larger number so the random uncertainty becomes smaller. This is the reason for carrying out repeated readings in experiments.

Systematic uncertainties are uncertainties to do with the setup of a system. For example, if a meter has not been zeroed correctly, each reading will be out by the same amount. This is a systematic uncertainty.

If a graph from an experiment is a straight line but does not quite pass through the origin, you should consider the possibility that you have a systematic uncertainty. If each reading was out by the same amount, this may be what is preventing the graph from passing through the origin.

Key points

* Use the correct units for all the quantities that you use in examples and questions.
* Understand what the prefixes before units mean and be able to change them into scientific notation.
* Take care when using significant figures. Answer questions to the same number of significant figures as used in the question.
* Be able to identify and calculate uncertainties in measurements.
* Understand the difference between absolute, percentage, random, scale-reading and systematic uncertainties.
* When taking a series of measurements there will be an uncertainty in each measurement. The largest percentage uncertainty in these results is used as the uncertainty in the final result.

Key words

Absolute uncertainty – the uncertainty in a measurement or the result of a calculation
Percentage uncertainty – the absolute uncertainty divided by the measurement multiplied by 100%
Random uncertainty – the uncertainty associated with repeated readings of the same quantity
Systematic uncertainty – an uncertainty caused by the system, such as a meter not being calibrated correctly

Questions

1 A stop clock is used to measure the time for ten swings of a pendulum. The experiment is repeated five times and the following results are obtained:

10·8 s, 10·4 s, 11·2 s, 10·7 s, 11·1 s
Calculate:
a) the average time for ten swings
b) the random uncertainty in the results.

2 A trolley is allowed to run through a light gate which is attached to a timer. The following results are obtained:

Length of trolley = 285 ± 1 mm
Time on timer = 0·42 ± 0·01 s
a) Calculate the percentage uncertainty in:
(i) the length of the trolley
(ii) the time on the timer.
b) Calculate the speed of the trolley. Express your answer as value ± absolute uncertainty.

Answers

Chapter 1

1. They are balanced.

2. $u = 2.0\,\mathrm{m\,s^{-1}}$
   $v = 8.0\,\mathrm{m\,s^{-1}}$
   $a = 1.5\,\mathrm{m\,s^{-2}}$
   $v^2 = u^2 + 2as$
   $8^2 = 2^2 + 2 \times 1.5 \times s$
   $64 = 4 + 3s$
   $60 = 3s$
   $s = 20\,\mathrm{m}$

3. $u = 12\,\mathrm{m\,s^{-1}}; v = 0\,\mathrm{m\,s^{-1}}; a = -4.0\,\mathrm{m\,s^{-2}}; v = u + at$
   $0 = 12 + (-4t)$
   $4t = 12$
   $t = 3\,\mathrm{s}$

4.

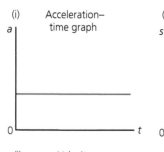

   (i) Acceleration–time graph (ii) Displacement–time graph

5.

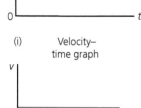

   (i) Velocity–time graph (ii) Acceleration–time graph

6. a) The acceleration is constant.
   b) Down
   c) $u = 0.0\,\mathrm{m\,s^{-1}}; t = 1.0\,\mathrm{s}; a = -9.8\,\mathrm{m\,s^{-2}};$
   $s = ut + \frac{1}{2}at^2$
   $s = (0 \times 1) + (\frac{1}{2} \times (-9.8) \times 1^2)$
   $s = -4.9\,\mathrm{m}$ (Note the displacement has a negative value as the ball has moved downwards.)

*7. a) (i) $d = vt = 20 \times 3.06 = 61.2\,\mathrm{m}$
       (ii) $v^2 = u^2 + 2as$
       $0 = 15^2 + 2 \times (-9.8) \times s$
       $s = 11.5\,\mathrm{m}$

b) More likely because: horizontal velocity will decrease; range will decrease; time in air will decrease; height reached will decrease

Chapter 2

1. a) Drag/friction; thrust from tail
   b) Drag/friction
   c) Component of weight; friction
   d) Thrust; drag

2. a) h = 122 N; v = 26 N
   b) h = 30 N; v = 63 N
   c) h = 498 N; v = 418 N

3. a) 294 J
   b) $7\,\mathrm{m\,s^{-1}}$
   c) (i) 71 J
      (ii) Heat (due to drag)

4. a) $4.8\,\mathrm{m\,s^{-1}}$
   b) 25 000 N

5. a) $2.2\,\mathrm{m\,s^{-1}}$
   b) 3.9 m
   c) 0.15 m

6. a) (i) $E_k = \frac{1}{2}mv^2 = \frac{1}{2} \times 0.24 \times 0.16^2 = 0.0031\,\mathrm{J}$
      (ii) The box is moving at a constant speed so the forces acting on it are balanced. The friction force will therefore be equal in size but opposite in direction to the component of weight down the slope.
      $F = mg \sin\theta = 0.24 \times 9.8 \times \sin 30° = 1.2\,\mathrm{N}$
   b) $E_W = Fd = 1.2 \times 4.2 = 5.0\,\mathrm{J}$

7. a) (i) $E_p = mgh = 2 \times 9.8 \times 1.6 = 31\,\mathrm{J}$
      (ii) The slope is frictionless so all of the potential energy will change into kinetic energy.
      $E_k = \frac{1}{2}mv^2$
      $v = \sqrt{\frac{2E_k}{m}} = \sqrt{2 \times \frac{31}{2}} = 5.6\,\mathrm{m\,s^{-1}}$
   b) This has no effect. The height remains the same so the potential energy remains the same. This means that the kinetic energy also remains the same so the speed must remain the same.

c) This has no effect. We use the fact that $E_k = \frac{1}{2}mv^2 = E_p = mgh$ to work out the speed. Because mass appears in both equations, they cancel each other out. Therefore changing the mass has no effect on the speed at the bottom of the slope.

8 a) $E_w = Fd = 20 \times 6 = 120\,J$

b) Because $\sin\theta =$ opposite/hypotenuse we can calculate the height of the ramp.
$h = \sin 30° \times$ length of ramp $= \frac{1}{2} \times 6$
$= 3.0\,m$
$E_p = mgh = 1.2 \times 9.8 \times 3 = 35\,J$

9 When the ball is rolling up the slope, the two forces acting on it are the component of weight down the slope and friction. As it is travelling up the slope, friction acts down the slope so both forces are acting in the downwards direction. When the ball starts to run down the slope, the component of weight is still acting downwards but the friction now acts up the slope (friction always acts opposite to the direction of travel). This means that the unbalanced force must now be smaller so the acceleration must also be smaller.

10 a) Lift = mg **or** lift = weight **or** forces balanced
$W = 9.75 \times 9.8 = 96.5\,N$

*b) Weight is less. There is a resultant force upwards **or** unbalanced force upwards **or** net force upwards.
Upward acceleration **or** The drone accelerates upwards.

*11 a) $v = u + at$
$20 = 0 + 4a$
$a = 5.0\,m\,s^{-2}$

b) car: $d = vt = 15 \times 4 = 60$
motorcycle: $s = ut + \frac{1}{2}at^2$
$= \frac{1}{2} \times 5 \times 16 = 40$
Extra distance $= 60 - 40 = 20\,m$

c) (i) $F_{(resultant)} = ma = 290 \times 5 = 1450(N)$
Frictional force $= 1450 - 1800$
$= (-)350\,N$

(ii) The faster it goes, the greater the air resistance **or** frictional forces/friction/drag. If $F_{(drive)}$ is constant, the unbalanced force would decrease **or** increasing $F_{(drive)}$ keeps the unbalanced force constant.

Chapter 3

1 $p = mv = 1200 \times 12 = 14\,400\,kg\,m\,s^{-1}$

2 Mass of ball A, $m_A = 2.8\,kg$
Mass of ball B, $m_B = 1.0\,kg$
Before the collision:
velocity of ball A, $u_A = 4.0\,m\,s^{-1}$
velocity of ball B, $u_B = 0.0\,m\,s^{-1}$
After the collision:
velocity of ball A, $v_A = v$
velocity of ball B, $v_B = 8.4\,m\,s^{-1}$
$$m_A u_A + m_B u_B = m_A v_A + m_B v_B$$
$$(2.8 \times 4.0) + (1.0 \times 0) = 2.8v + (1.0 \times 8.4)$$
$$11.2 + 0 = 2.8v + 8.4$$
$$11.2 - 8.4 = 2.8v$$
$$2.8 = 2.8v$$
$$v = 1.0\,m\,s^{-1}$$

3 Total momentum before $= 0$
$$m_A u_A + m_B u_B = m_A v_A + m_B v_B$$
$$0 = 2.2v + (0.2 \times 8)$$
$$-1.6 = 2.2v$$
$$v = -0.73\,m\,s^{-1}$$
The recoil velocity is negative because it is in the opposite direction.

4 a) $0.2\,m\,s^{-1}$ to the right
b) E_k before $= 2.4 + 2.56 = 4.96\,J$; E_k after $= 0.6 + 0.04 = 0.64\,J$: collision is inelastic
c) 24 N to the left
d) 24 N to the right

5 a) Impulse = area under the graph
$$= \frac{1}{2} \times 0.075 \times 120$$
$$= 4.5\,N\,s$$
b) Change in momentum $= 4.5\,N\,s$
Change in momentum $= m(v - u)$
$$4.5 = 1.5(v - (-5.0))$$
$$4.5 = 1.5v + 7.5$$
$$-3.0 = 1.5v$$
$$v = -2.0\,m\,s^{-1}$$
The velocity is negative because the ball is now travelling in the opposite direction.
c) The graph would have a lower peak and the time of contact would be longer. The foam changes shape so the time of contact is longer and so the maximum force will be less.

6 D

7 C

8 E

Chapter 4

1. a) $21.3\,\mathrm{m\,s^{-1}}$; $5.7\,\mathrm{m\,s^{-1}}$
   b) $0.58\,\mathrm{s}$
   c) $1.7\,\mathrm{m}$
   d) $24.8\,\mathrm{m}$

2. a) $1.8\,\mathrm{m}$
   b) The time it takes to fall is independent of the velocity of the ball. The time is determined by the height the ball falls from. It will take the same time to fall but as it is travelling at twice the horizontal velocity it will cover twice the distance.

3. a) $u_h = u\cos\theta = 12 \times \cos 60° = 6.0\,\mathrm{m\,s^{-1}}$
   b) $u_v = u\sin\theta = 12 \times \sin 60° = 10.4\,\mathrm{m\,s^{-1}}$
   c) At the maximum height, $v_v = 0$
   $t = \dfrac{v-u}{a} = \dfrac{0-10.4}{-9.8} = 1.1\,\mathrm{s}$
   d) $s = ut + \frac{1}{2}at^2 = (10.4 \times 1.1) + (\frac{1}{2} \times (-9.8) \times 1.1^2) = 11.44 - 5.93 = 5.51\,\mathrm{m}$
   e) $13.2\,\mathrm{m}$

4. The satellite is following a curved path so its velocity is changing. If the velocity is changing, the satellite must be accelerating.

5. When the projectile is travelling at an angle of 45° then its vertical and horizontal components must be the same size. We can use Pythagoras' Theorem to calculate the horizontal velocity at this time.
   $a^2 + b^2 = c^2$
   but $a = b$
   so $2a^2 = c^2$
   $2a^2 = 100$
   $a^2 = 50$
   $a = 7.1\,\mathrm{m\,s^{-1}}$
   The horizontal velocity does not change so the initial horizontal velocity was $7.1\,\mathrm{m\,s^{-1}}$.

6. a) $F = G\dfrac{m_1 \times m_2}{r^2}$
   $= 6.67 \times 10^{-11} \times \dfrac{(0.1 \times 10^{-3}) \times (0.1 \times 10^{-3})}{1^2}$
   $= 6.67 \times 10^{-19}\,\mathrm{N}$

   b) $F = G\dfrac{m_1 \times m_2}{r^2}$
   $= 6.67 \times 10^{-11} \times \dfrac{(0.1 \times 10^{-3}) \times (0.1 \times 10^{-3})}{0.001^2}$
   $= 6.67 \times 10^{-13}\,\mathrm{N}$

7. $F = G\dfrac{m_1 \times m_2}{r^2}$
   $= 6.67 \times 10^{-11} \times \dfrac{(6.0 \times 10^{24}) \times (7.3 \times 10^{22})}{\left(3.84 \times 10^8\right)^2}$
   $= 1.98 \times 10^{20}\,\mathrm{N}$

8. The gravitational field strength is the force on a 1 kg mass:
   $F = G\dfrac{m_1 \times m_2}{r^2} = 6.67 \times 10^{-11} \times \dfrac{1 \times (7.3 \times 10^{22})}{\left(1.7 \times 10^6\right)^2}$
   $= 1.68\,\mathrm{N}$
   The gravitational field strength is therefore $1.68\,\mathrm{N\,kg^{-1}}$.

9. $r^2 = G\dfrac{m_1 \times m_2}{F} = 6.67 \times 10^{-11} \times \dfrac{200 \times (6 \times 10^{24})}{200}$
   $= 4 \times 10^{14}$
   $r = 2.0 \times 10^7\,\mathrm{m}$
   This is the radius of the orbit so the height above the surface of the Earth is $2.0 \times 10^7\,\mathrm{m}$ $- 6.4 \times 10^6\,\mathrm{m} = 13.6 \times 10^6\,\mathrm{m}$.

*10 a) (i) $u_h = 6.5\cos 50° = 4.2\,\mathrm{m\,s^{-1}}$
   (ii) $u_v = 6.5\sin 50° = 5.0\,\mathrm{m\,s^{-1}}$
   b) $t = \dfrac{s}{v} = \dfrac{2.9}{4.2} = 0.69\,(\mathrm{s})$
   c) $s = ut + \frac{1}{2}at^2 = 5 \times 0.69 + \frac{1}{2} \times (-9.8) \times (0.69)^2 = 1.1\,\mathrm{m}$
   so height $h = 2.3 + 1.1 = 3.4\,\mathrm{m}$
   d) The ball would **not** land in basket; (initial) vertical speed would increase; so the ball is higher than the basket when it has travelled 2.9 m horizontally **or** So the ball has travelled further horizontally when it is at the same height as the basket.

*11 a) (i) $v^2 = u^2 + 2as$
   $0 = 7^2 + 2 \times (-9.8) \times s$
   $s = 2.5\,\mathrm{m}$
   or
   $v = u + at$
   $0 = 7 + (-9.8)t$
   $t = 0.71\,\mathrm{s}$
   $s = ut + \frac{1}{2}at^2$
   $= 7 \times 0.71 + \frac{1}{2}(-9.8)(0.71)^2$
   $= 2.5\,\mathrm{m}$
   (ii) $v = u + at$
   $0 = 7 + (-9.8) \times t$
   $t = 0.71\,\mathrm{s}$
   or
   $s = \frac{1}{2}(u + v)t$
   $2.5 = \frac{1}{2}(7 + 0)t$
   $t = 0.71\,\mathrm{s}$

b) (i) $1.5\,\text{m s}^{-1}$ to the **right**

   (ii) Statement Z: **Horizontal** speed of the ball remains constant and equal to the (horizontal) speed of the trolley **or** Horizontal speed of the ball remains constant at $1.5\,\text{m s}^{-1}$.

Chapter 5

1 $t' = \dfrac{t}{\sqrt{1-\left(\frac{v}{c}\right)^2}}$

$t' = 10 \div \sqrt{1-0.2^2}$

$ = 10 \div \sqrt{1-0.04}$

$ = 10 \div \sqrt{0.96}$

$ = 10 \div 0.9798$

$ = 10.2\,\text{s}$

2 a) $t = s \div v$

$ = 2.1 \times 10^{11} \div 3.0 \times 10^7$

$ = 7.0 \times 10^3\,\text{s}$

  b) $t' = \dfrac{t}{\sqrt{1-\left(\frac{v}{c}\right)^2}}$

$t = t' \times \sqrt{(1-(v \div c)^2)}$

$ = 7.0 \times 10^3 \times \sqrt{(1-0.1^2)}$

$ = 7.0 \times 10^3 \times \sqrt{(1-0.01)}$

$ = 7.0 \times 10^3 \times \sqrt{(0.99)}$

$ = 7.0 \times 10^3 \times 0.995$

$ = 6.97 \times 10^3\,\text{s}$

  c) $l' = l\sqrt{1-\left(\frac{v}{c}\right)^2}$

$ = 20.0 \times \sqrt{(1-0.1^2)}$

$ = 20.0 \times \sqrt{(1-0.01)}$

$ = 20.0 \times \sqrt{(0.99)}$

$ = 20.0 \times 0.99$

$ = 19.9\,\text{m}$

Chapter 6

1 a) $f_o = f_s \times \dfrac{v}{(v-vs)} = 800\dfrac{340}{340-40}$

$ = 800 \times \dfrac{340}{300} = 800 \times 1.13 = 907\,\text{Hz}$

  b) $f_o = f_s \times \dfrac{v}{(v-vs)} = 800 \times \dfrac{340}{340+50} = 800 \times \dfrac{340}{390}$

$ = 800 \times 0.872 = 697\,\text{Hz}$

2 a) $v = zc = 0.0034 \times 3 \times 10^8 = 1.02 \times 10^6\,\text{m s}^{-1}$

  b) $H_o = \dfrac{v}{d} = \dfrac{1.02 \times 10^6}{4.44 \times 10^{23}} = 2.30 \times 10^{-18}\,\text{s}^{-1}$

  c) $t = \dfrac{1}{H_o} = \dfrac{1}{2.3 \times 10^{-18}} = 4.35 \times 10^{17}\,\text{s}$

$ = \dfrac{4.35 \times 10^{17}}{365 \times 24 \times 60 \times 60} = 13.8 \times 10^9\,\text{years}$

$ = 13.8\,\text{billion years}$

Chapter 7

1 Blue dwarfs

2 It increases

3 It has decreased

4 The abundance of hydrogen and helium in the Universe; the cosmic microwave background radiation; the cosmological red shift

Chapter 8

1 a) $E = QV = 1.6 \times 10^{-19} \times 5000 = 8.0 \times 10^{-16}\,\text{J}$

  b) $E_k = 8.0 \times 10^{-16}\,\text{J}$

$v = \sqrt{\dfrac{2E_k}{m}} = \sqrt{\dfrac{2 \times 8.0 \times 10^{-16}}{9.11 \times 10^{-31}}} = \sqrt{1.756 \times 10^{15}} = 4.2 \times 10^7\,\text{ms}^{-1}$

  c) Charge is the same, V is the same so E_k is the same.

$v = \sqrt{\dfrac{2E_k}{m}} = \sqrt{\dfrac{2 \times 8.0 \times 10^{-16}}{1.67 \times 10^{-27}}} = \sqrt{9.58 \times 10^{11}} = 9.8 \times 10^5\,\text{ms}^{-1}$

2 a) (i) The charge on an electron is the same size as the charge on a proton. The kinetic energy gained is equal to the work done by the electric field.

$E = QV = 1.6 \times 10^{-19} \times 2000 = 3.2 \times 10^{-16}\,\text{J}$

  (ii) $v = \sqrt{\dfrac{2E_k}{m}} = \sqrt{\dfrac{2 \times 3.2 \times 10^{-16}}{1.67 \times 10^{-27}}} = \sqrt{3.83 \times 10^{11}}$

$ = 6.19 \times 10^5\,\text{m s}^{-1}$

  b) (i) It will be the same. The potential difference is the same and the size of the charge is the same, so the work done by the field will be the same, so the kinetic energy will be the same.

  (ii) The velocity will be greater. The kinetic energy is the same, but the mass of the electron is smaller, so the velocity must be greater.

3 Advantage – much higher energies can be reached. Disadvantage – synchrotron radiation is emitted causing particles to lose energy.

4 A

5 a) $E_w = QV = 1.6 \times 10^{-19} \times 2250 = 3.6 \times 10^{-16}\,\text{J}$

  b) Work done $= \frac{1}{2}mv^2 = \frac{1}{2} \times 2.18 \times 10^{-25} \times v^2$

$v = 5.75 \times 10^4\,\text{m s}^{-1}$

*6 E

Chapter 9

1 Leptons or fermions
2 6
3 The study of beta decay
4 Two down-quarks and one up-quark
5 Two anti-up-quarks and one anti-down-quark
6 Neutral mesons are made up of two quarks. For the charge to be an integer number they must be made of one quark and one antiquark. This pair of matter–antimatter particles will in a very short time annihilate themselves.
7 Strong – gluon; electromagnetic – photon; weak – W^+, W^-, Z bosons; gravity – graviton

Chapter 10

1 a) $A = 0$, $B = -1$; beta decay
   b) Neutrino
2 a) X_1 = uranium, X_5 = thorium
   b) 2
   c) 2
3 The difference in mass between the left-hand side and the right-hand side can be found using $E = mc^2$. Rearranging gives $m = E \div c^2$
   $m = 2.97 \times 10^{-12} \div (3 \times 10^8)^2$
   $m = 0.033 \times 10^{-27}\,kg$
   The total mass of the right-hand side is
   $$2 \times 6.642 \times 10^{-27} \quad\quad 13.284 \times 10^{-27}$$
   $$+ \quad\quad\quad\quad\quad\quad\quad\quad 1.675 \times 10^{-27}$$
   $$\overline{\quad\quad\quad\quad\quad\quad\quad\quad 14.959 \times 10^{-27}\,kg}$$
   The total mass of the left-hand side is
   $$14.959 \times 10^{-27}$$
   $$+ \quad\; 0.033 \times 10^{-27}$$
   $$\overline{14.992 \times 10^{-27}\,kg}$$
   The mass of the $^7_3 Li$ isotope is
   $$14.992 \times 10^{-27}$$
   $$- \quad\; 3.342 \times 10^{-27}$$
   $$\overline{11.650 \times 10^{-27}\,kg}$$
4 a) Induced
   b) The total mass of the left-hand side is
   $$390.173 \times 10^{-27}$$
   $$+ \quad\quad 1.675 \times 10^{-27}$$
   $$\overline{391.848 \times 10^{-27}\,kg}$$
   The total mass of the right-hand side is
   $$230.584 \times 10^{-27}$$
   $$157.544 \times 10^{-27}$$
   $$+ \; 2 \times 1.675 \times 10^{-27} \quad\quad 3.350 \times 10^{-27}$$
   $$\overline{391.478 \times 10^{-27}\,kg}$$

The difference in mass is
$$391.848 \times 10^{-27}$$
$$- \;\; 391.478 \times 10^{-27}$$
$$\overline{0.37 \times 10^{-27}\,kg}$$
The energy released is therefore
$E = mc^2$
$E = 0.37 \times 10^{-27} \times (3 \times 10^8)^2$
$E = 3.33 \times 10^{-11}\,J$

*5 a) (Nuclear) fusion
   b) Total mass before = $3.342 \times 10^{-27} + 5.005 \times 10^{-27} = 8.347 \times 10^{-27}\,kg$
   Total mass after = $6.642 \times 10^{-27} + 1.675 \times 10^{-27} = 8.317 \times 10^{-27}\,kg$
   Loss in mass = $0.030 \times 10^{-27}\,kg$
   Energy released = mc^2
   $0.030 \times 10^{-27} \times (3.00 \times 10^8)^2 = 2.7 \times 10^{-12}\,J$

6 A
*7 B

Chapter 11

1 Coherent waves have the same frequency, wavelength and speed and they have a constant phase relationship.
2 When a crest from one source meets a crest from another source (or a trough from one source meets a trough from another source) they produce a double height crest (or double depth trough).

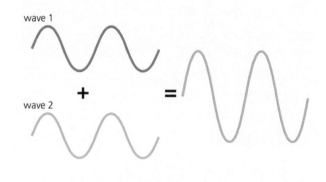

3 a) At the first maximum, the path difference is 1λ.
   $\lambda = 1.43 - 0.75 = 0.68\,m$
   b) $v = f\lambda$
   $f = \frac{v}{\lambda} = \frac{340}{0.68} = 500\,Hz$
4 d = distance between two slits
   $= \frac{1}{400000}$
   $= 2.5 \times 10^{-6}\,m$
   $m\lambda = d\,sin\,\theta$
   $2 \times \lambda = 2.5 \times 10^{-6} \times sin\,25°$
   $2 \times \lambda = 2.5 \times 10^{-6} \times 0.423$
   $\lambda = 528\,nm$

5 A

6 C

Chapter 12

1 a) Nothing. The energy of each photon is less than the work function of the metal so no electrons will be emitted.

   b) Nothing. Increasing the wavelength decreases the frequency of the photons. As they are already below the threshold frequency, no electrons will be emitted.

2 a) $f_o = \dfrac{E}{h} = \dfrac{2 \cdot 65 \times 10^{-18}}{6 \cdot 63 \times 10^{-34}} = 4 \cdot 0 \times 10^{15}\,\text{Hz}$

   b) $E = hf = 6 \cdot 63 \times 10^{-34} \times 5 \cdot 6 \times 10^{15} = 3 \cdot 71 \times 10^{-18}\,\text{J}$
   $E_k = E - W = 3 \cdot 71 \times 10^{-18} - 2 \cdot 65 \times 10^{-18}$
   $= 1 \cdot 06 \times 10^{-18}\,\text{J}$

3 a) $f = \dfrac{v}{\lambda} = \dfrac{3 \times 10^{8}}{167 \times 10^{-9}} = 1 \cdot 80 \times 10^{15}\,\text{Hz}$

   b) $E = hf = 6 \cdot 63 \times 10^{-34} \times 1 \cdot 80 \times 10^{15} = 1 \cdot 19 \times 10^{-18}\,\text{J}$

   c) $W = hf_o = 6 \cdot 63 \times 10^{-34} \times 1 \cdot 2 \times 10^{15}$
   $= 7 \cdot 96 \times 10^{-19}\,\text{J}$
   $E_k = E - W = 1 \cdot 19 \times 10^{-18} - 7 \cdot 96 \times 10^{-19}$
   $= 3 \cdot 94 \times 10^{-19}\,\text{J}$

   d) $v = \sqrt{\dfrac{2E_k}{m}} = \sqrt{\dfrac{2 \times 3 \cdot 94 \times 10^{-19}}{9 \cdot 11 \times 10^{-31}}} = \sqrt{8 \cdot 64 \times 10^{11}}$
   $= 9 \cdot 3 \times 10^{5}\,\text{m s}^{-1}$

4 Irradiance is the power per unit area and is measured in W m^{-2}.

5 $P = \dfrac{E}{t} = \dfrac{1200}{60} = 20\,\text{W}$
   $I = \dfrac{P}{A} = \dfrac{20}{4} = 5 \cdot 0\,\text{W m}^{-2}$

6 $I = \dfrac{k}{d^2}$
   $k = Id^2$
   $= 64 \times 1 \cdot 25^2 = 100$
   $I = \dfrac{k}{d^2} = \dfrac{100}{2 \cdot 5^2} = 16\,\text{W m}^{-2}$

7

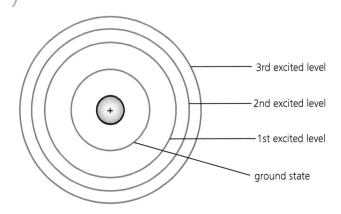

- 3rd excited level
- 2nd excited level
- 1st excited level
- ground state

8 a) (i) 6 (B $\rightarrow$ C, C $\rightarrow$ D, D $\rightarrow$ ground, B $\rightarrow$ D, C $\rightarrow$ ground, B $\rightarrow$ ground)

      (ii) 6

   b) (i) $E = -1 \cdot 80 \times 10^{-19} - (-6 \cdot 23 \times 10^{-19})$
   $= 4 \cdot 43 \times 10^{-19}\,\text{J}$

      (ii) $E = hf$
   $f = \dfrac{E}{h} = \dfrac{4 \cdot 43 \times 10^{-19}}{6 \cdot 63 \times 10^{-34}} = 6 \cdot 68 \times 10^{14}\,\text{Hz}$

   c) The longest wavelength will be the lowest frequency, so it will be the smallest energy transition, B $\rightarrow$ C.
   $E = -4 \cdot 00 \times 10^{-20} - (-1 \cdot 80 \times 10^{-19})$
   $= 1 \cdot 40 \times 10^{-19}\,\text{J}$
   $f = \dfrac{E}{h} = \dfrac{1 \cdot 40 \times 10^{-19}}{6 \cdot 63 \times 10^{-34}} = 2 \cdot 11 \times 10^{14}\,\text{Hz}$
   $\lambda = \dfrac{v}{f} = \dfrac{3 \cdot 0 \times 10^{8}}{2 \cdot 11 \times 10^{14}} = 1 \cdot 42 \times 10^{-6}\,\text{m}$

9 a) Energy released $= -2.419 \times 10^{-19} - (-21.760 \times 10^{-19}) = 19.341 \times 10^{-19}\,\text{J}$
   $E = hf$
   $19.341 \times 10^{-19} = 6 \cdot 63 \times 10^{-34} \times f$
   $f = 2 \cdot 92 \times 10^{15}\,\text{Hz}$
   $\lambda = \dfrac{c}{f} \qquad = \dfrac{3 \times 10^{8}}{2.92 \times 10^{15}}$
   $= 102\,\text{nm}$

   b) 'Blue' **or** 'blue/green'

10 a) (i) $E_k = hf - hf_o = 5 \cdot 95 \times 10^{-19} - 2 \cdot 94 \times 10^{-19}$
   $= 3 \cdot 01 \times 10^{-19}\,\text{J}$

      (ii) $E_k = \tfrac{1}{2}mv^2$
   $3.01 \times 10^{-19} = \tfrac{1}{2} \times 9 \cdot 11 \times 10^{-31} \times v^2$
   $v = 8 \cdot 13 \times 10^{5}\,\text{m s}^{-1}$

   b) There will be no change to the velocity. More photons will eject more electrons but the energy of the photons is still the same.

Chapter 13

1 a) $44 \cdot 8^{\circ}$

   b) $36 \cdot 8^{\circ}$

   c) $22 \cdot 6^{\circ}$

2 The angle between the normal and the ray in air $= 90 - 38 = 52^{\circ}$. The angle between the normal and the ray in the material $= 90 - 53 = 37^{\circ}$.
   $n = \dfrac{\sin \theta_{air}}{\sin \theta_{material}} = \dfrac{\sin 52}{\sin 37} = 1 \cdot 31$

3 a) Increases

b) Increases

c) Increases

d) Stays the same

4 White light is made up of a range of frequencies. The refractive index increases with frequency. Blue light will be deviated more when it enters the block than red. The sides of the prism slope in opposite directions so when the light leaves the prism the colours are spread further apart.

5 a) $v_{ice} = \frac{v_{air}}{n} = \frac{3 \times 10^8}{1{\cdot}31} = 2{\cdot}29 \times 10^8\,\text{m s}^{-1}$

b) $\sin \square_c = \frac{1}{n} = \frac{1}{1{\cdot}31} = 0{\cdot}763$

$\theta_c = 49{\cdot}7\,^\circ$

6 a) $\frac{v_{air}}{v_{glass}} = \frac{\sin \theta_{air}}{\sin \theta_{glass}}$

$v_{glass} = \frac{v_{air} \times \sin \theta_{glass}}{\sin \theta_{air}} = \frac{3 \times 10^8 \times \sin 48{\cdot}9}{\sin 60}$

$= 2{\cdot}61 \times 10^8\,\text{m s}^{-1}$

b) $\frac{v_{air}}{v_{glass}} = \frac{\lambda_{air}}{\lambda_{glass}}$

$\lambda_{glass} = \frac{\lambda_{air} \times v_{glass}}{v_{air}} = \frac{640 \times 10^{-9} \times 2{\cdot}61 \times 10^8}{3 \times 10^8}$

$= 557 \times 10^{-9}\,\text{m}$

7 $f = \frac{v}{\lambda} = \frac{3 \times 10^8}{4{\cdot}75 \times 10^{-7}} = 6{\cdot}3 \times 10^{-7}\,\text{Hz}$

*8 B

*9 a) $n = \frac{\sin \theta_1}{\sin \theta_2}$

$1{\cdot}49 = \frac{\sin \theta_{air}}{\sin 19}$

$\theta_{air} = 29\,^\circ$

b) $n = \frac{1}{\sin \theta_c}$

$1{\cdot}49 = \frac{1}{\sin \theta_c}$

$\theta_c = 42\,^\circ$

c) Different frequencies/colours are <u>refracted</u> through different angles **or** The <u>refractive index</u> is different for different frequencies/colours.

*10 a) $n = \frac{\sin \theta_1}{\sin \theta_2}$

$1{\cdot}33 = \frac{\sin X}{\sin 36}$

$X = 51\,^\circ$

b) (i) Angle of refraction is 90° **or** Refracted ray makes an angle of 90° with normal **or** Refracted ray is along surface of water

(ii) $\sin \theta_c = \frac{1}{n} = \frac{1}{1{\cdot}33}$

$\theta_c = 49\,^\circ$

c) Totally internally reflected ray shown (angles should be equal)

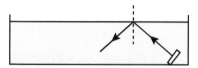

Chapter 14

1 a) 50 Hz means that there are 50 cycles per second

b) $T = \frac{1}{50} = 0{\cdot}02\,\text{s}$

c) $V_{peak} = V_{r.m.s.} \times \sqrt{2} = 230 \times \sqrt{2} = 325\,\text{V}$

d) $I = V \div R = \frac{230}{57{\cdot}5} = 4{\cdot}0\,\text{A}$

e) $I_{peak} = I_{r.m.s.} \times \sqrt{2} = 4{\cdot}0 \times \sqrt{2} = 5{\cdot}7\,\text{A}$

2 a) Time for one wave (period) $= 8 \times 0{\cdot}1 \times 10^{-3}$

$= 0{\cdot}8 \times 10^{-3}\,\text{s}$

$f = \frac{1}{T} = \frac{1}{0{\cdot}0008} = 1250\,\text{Hz}$

b) $V_{peak} = 0{\cdot}5 \times 3 = 1{\cdot}5\,\text{V}$

c) $V_{peak} = \frac{V_p}{\sqrt{2}} = \frac{1{\cdot}5}{\sqrt{2}} = 1{\cdot}1\,\text{V}$

3 a) The mains voltage is 230 V.

$P = IV = 230 \times 12 = 2760\,\text{W}$

b) $R = \frac{V}{I} = \frac{230}{12} = 19{\cdot}2\,\Omega$

4 a) (i) Connect all three in series.

(ii) Connect all three in parallel.

b) Calculate the resistance of 12 Ω and 24 Ω in parallel:

$\frac{1}{R_T} = \frac{1}{R_1} + \frac{1}{R_2} = \frac{1}{12} + \frac{1}{24} = \frac{2}{24} + \frac{1}{24} = \frac{3}{24}$

$R_T = \frac{24}{3} = 8\,\Omega$

If this is connected in series to the 48 Ω, the total resistance will be 56 Ω. Therefore the 12 Ω and 24 Ω resistors should be connected in parallel and this network should then be connected in series with the 48 Ω resistor.

5 a) $V_1 = \frac{R_1}{(R_1 + R_2)} \times V_s = \frac{100}{(100 + 250)} \times 5 = \frac{100}{350} \times 5$

$= 1{\cdot}43\,\text{V}$

b) $V_2 = V_s - V_1 = 5 - 1{\cdot}43 = 3{\cdot}57\,\text{V}$

$P = \frac{V^2}{R} = \frac{3{\cdot}57^2}{250} = 0{\cdot}051\,\text{W}$

6   $\frac{1}{R_p} = \frac{1}{R_1} + \frac{1}{R_2} = \frac{1}{60} + \frac{1}{60} = \frac{2}{60}$

   $R_p = \frac{60}{2} = 30\,\Omega$

   We can now solve the problem as if it was a voltage divider.

   $V_1 = \frac{R_1}{(R_1 + R_2)} \times V_s = \frac{16}{(16 + 30)} \times 230 = \frac{16}{46} \times 230 = 80\,V$

7   The energy converted per second is the power.

   $R_T = R_1 + R_2 + R_3 = 150 + 350 + 1000 = 1500\,\Omega$

   $P = \frac{V^2}{R} = \frac{6^2}{1500} = 0.024\,W$

8   The EMF is the number of joules of energy given to each coulomb of charge that passes through the source.

9   Zero ohms

10 a) Total resistance $= 1.8 + 0.4 = 2.2\,\Omega$

   $I = V \div R = 4.4 \div 2.2 = 2.0\,A$

   b) $V = IR = 2 \times 1.8 = 3.6\,V$

   c) $V_L = E - V = 4.4 - 3.6 = 0.8\,V$

11 $I = E \div r = 1.5 \div 0.3 = 5.0\,A$

12 a)

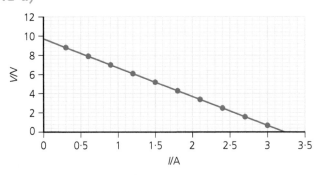

   b) $E = $ intercept of graph $= 9.7\,V$

   c) $r = \frac{E-V}{I} = \frac{9.7-6.1}{1.2} = \frac{3.6}{1.2} = 3.0\,\Omega$

13 a) $C = \frac{Q}{V} = \frac{42 \times 10^{-9}}{12} = 3.5 \times 10^{-9}\,F$

   b) $E = \frac{1}{2}QV = \frac{1}{2} \times 42 \times 10^{-9} \times 12 = 252 \times 10^{-9}\,J$

14

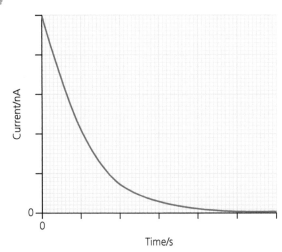

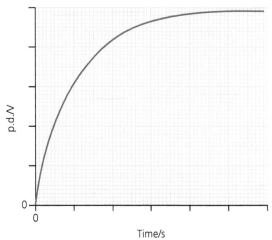

15 a) $I = \frac{V}{R} = \frac{24}{47000} = 0.511\,mA$

   b) $Q = CV = 24 \times 50 \times 10^{-6} = 1.2\,mC$

   c) When the voltage across the resistor is 8.0 V, then the voltage across the capacitor is 16.0 V.

   $E = \frac{1}{2}CV^2 = \frac{1}{2} \times 50 \times 10^{-6} \times 16^2 = 6.4\,mJ$

*16 a) $V_p = 3 \times 0.5 = 1.5\,mV$

   b) $f = \frac{1}{T} = \frac{1}{4 \times 10^{-3}} = 250\,Hz$

*17 a) (i) $V_p = 1.5\,V$

   (ii) $f = \frac{1}{T} = \frac{1}{0.008} = 125\,Hz$

   b) Stays the same/constant/no change/nothing

*18 a) (i) 0.22 V

   (ii) $E = V + Ir$

   $0.22 = 0.10 + 3r$

   $r = 0.04\,\Omega$

   or

   $E = I(R + r)$

   $0.22 = 0.5(0.4 + r)$

   $r = 0.04\,\Omega$

   b) $E = I(R + r)$

   $0.88 = I(0.12 + 0.15)$

   $I = 3.26\,A$

   Yes, because the current at 800 °C is 3.26 A, higher than the 2.5 A point where the gas valve closes.

*19 a) 200 μC of charge increases voltage across plates by 1 volt **or** 200 μC per volt **or** One volt across the plates of the capacitor causes 200 μC of charged to be stored.

   b) (i) $I = \frac{E}{R} = \frac{12}{1400} = 0.0086\,A\ (8.6\,mA)$

   (ii) $E = \frac{1}{2}CV^2$

   Initial stored energy $= \frac{1}{2} \times (200 \times 10^{-6}) \times 12^2$

   $= 0.0144\,J$

   Final stored energy $= \frac{1}{2} \times (200 \times 10^{-6}) \times 4^2$

   $= 0.0016\,J$

Difference = 0·0144 − 0·0016
Decrease in stored energy = 0·0128 J

c) (i) 0·30 s

(ii) $s = ut + \frac{1}{2}at^2$
$0·80 = 1·5 \times 0·3 + \frac{1}{2} \times a \times (0·3)^2$
$a = 7·8\,\mathrm{m\,s^{-2}}$

(iii) Percentage (fractional) uncertainty in (measuring) <u>distance</u> will be smaller **or** Percentage (fractional) uncertainty in (measuring) <u>time</u> will be smaller.

Chapter 15

1 a)

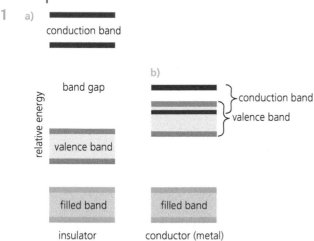

2 The resistance decreases
3 The conductivity increases so the resistance must decrease.

4

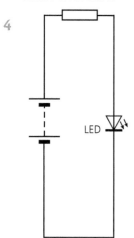

The resistor is used to limit the current in the LED.

Uncertainties

1 a) Average time =
$= \frac{total}{number} = \frac{10·8 + 10·4 + 11·2 + 10·7 + 11·1}{5}$
$= 10·84 = 10·8\,\mathrm{s}$

b) Random uncertainty =
$= \frac{max\ reading - min\ reading}{number\ or\ readings} = \frac{11·2 - 10·4}{5} = \frac{0·8}{5}$
$= 0·16 = 0·2\,\mathrm{s}$

2 a) (i) $\frac{1}{285} \times 100 = 0·35\%$

(ii) $\frac{0·01}{0·42} \times 100 = 2·4\%$

b) $v = \frac{d}{t} = \frac{0·285}{0·42} = 0·678571428 = 0·68\,\mathrm{m\,s^{-1}}$

The largest % uncertainty is 2·4% so this is the % uncertainty in the final result. The absolute uncertainty is 2·4% of 0·68 = 0·01632 = 0·02 m s⁻¹. Final answer is $v = 0·68 \pm 0·02\,\mathrm{m\,s^{-1}}$.

Glossary

Absolute uncertainty (p.137) – the uncertainty in a measurement or the result of a calculation

Absorption spectrum (p.91) – black lines in a continuous spectrum caused when the vapour of an element absorbs particular photons of light. The photons absorbed have exactly the same energy as energy gaps in the energy level within the atom of the element

AC (p.105) – alternating current

Acceleration (p.1) – the rate of change of velocity

Alpha (p.67) – a type of radioactive decay; when an isotope undergoes alpha decay, it releases a particle made up of two protons and two neutrons

Antimatter (p.63) – material made up of antiquarks and antileptons. Each matter particle has an antimatter particle equivalent with an opposite charge but identical mass

Balanced forces (p.7) – equivalent to having no force acting on an object; the forces cancel each other out

Baryon (p.64) – a particle made up of three quarks

Beta (p.64) – a type of radioactive decay; during beta decay a neutron in an isotope decays to become a proton, releasing a beta particle (an electron) and a neutrino

Bohr model (p.89) – a model of the atom with a central positive nucleus and electrons in definite energy levels surrounding the nucleus

Boson (p.64) – a force-mediating particle or a meson

Capacitance (p.115) – the ratio of charge to voltage for a capacitor; the number of coulombs of charge a capacitor can store per volt

Capacitor (p.114) – an electronic device that can store charge and energy in a DC circuit

Coherence (p.76) – coherent waves (or waves that have coherence) have the same speed, wavelength and frequency and have a constant phase relationship

Component (p.13) – part of something

Conduction band (p.130) – the band that allows conduction to take place; in conductors this band is partly filled

Conductor (p.130) – a material that allows electricity to flow

Conservation (p.15) – the total amount of a quantity stays the same

Constant (p.2) – not changing; uniform

Constructive (p.77) – two waves combining to produce a maximum

Cosmic microwave background radiation (p.50) – the observed background radiation that is present in every direction in the Universe; it provides strong evidence to support the Big Bang Theory of the Universe

Critical angle (p.99) – the angle of incidence in the material that causes a ray of light to refract into air at 90°

Current (p.105) – a flow of charge (or electrons)

Dark energy (p.49) – a type of energy proposed to account for the missing energy in the Universe; this is to account for the fact that the rate of expansion of the Universe is increasing

Dark matter (p.49) – a type of matter proposed to account for the missing mass of the Universe; this mass cannot be observed by conventional telescopes

DC (p.105) – direct current

Destructive (p.77) – two waves combining to produce a minimum

Diffraction (p.75) – the spreading of waves on passing an edge or through a narrow aperture

Dispersion (p.98) – visible light 'splitting' into various colours

Displacement (p.2) – how far an object is from a point in a particular direction

Doping (p.132) – the adding of impurities to pure silicon to produce n-type or p-type silicon

Doppler Effect (p.44) – a naturally occurring phenomenon which alters the relative frequency of radiation that we can detect

Elastic (p.24) – in collisions, an elastic collision is one in which kinetic energy is conserved

Electric field (p.52) – the region of space around a charge

Electric field pattern (p.52) – a pattern of lines around a charged object; the lines indicate the direction a positively charged particle will move in the field

Electromotive force (EMF) (p.111) – the number of joules of electrical energy given to each coulomb of charge that passes through the source

Emission spectrum (p.91) – lines of colour in a continuous spectrum associated with a particular electron transition between two energy levels within an atom

Energy level (p.89) – an electron orbit inside the Bohr atom

Explosion (p.24) – two or more objects move apart during an explosion; the kinetic energy gained comes from energy stored in a spring or a chemical

Fermion (p.63) – a matter particle

Ferrous (p.55) – containing or consisting of iron with a valency of two

Fission (p.68) – the splitting of a nucleus into two smaller nuclei

Frame of reference (p.39) – the background against which measurements are made

Frequency (p.44) – the number of times something happens per second; for example, the number of complete AC cycles per second or the number of waves per second

Friction (p.11) – the force that opposes the motion of all objects

Fusion (p.69) – the joining of two small nuclei to form a larger nucleus

Galaxy (p.46) – a large collection of stars

Gamma (p.67) – photons of electromagnetic energy released by a nucleus

Grating (p.79) – a sheet of transparent material with fine lines etched on it to produce diffraction; each gap in the grating acts as a coherent source

Gravitational field strength (p.34) – the weight per unit mass

Ground state (p.89) – the lowest permitted energy level in an atom

Hadron (p.64) – a particle made up of quarks

Hubble's constant (p.46) – Hubble's constant is the recessional velocity of an object divided by the distance from the Earth

Hubble's Law (p.46) – this law states that the further away from the Earth an object is, the faster it is moving away from the Earth; this law only applies to very distant objects

Ideal supply (p.111) – a supply that has no internal resistance

Impulse (p.25) – the average force multiplied by the time

Induced (p.68) – caused to happen; in the case of nuclear fission the addition of a neutron to a large nucleus, causing it to split into two smaller nuclei

Inelastic (p.24) – in collisions, an inelastic collision is one in which kinetic energy is lost

Insulator (p.130) – a material that blocks the flow of electricity

Interference (p.76) – a series of maxima and minima produced when waves meet

Internal resistance (p.111) – the resistance of the material that makes up the cell

Invariance (p.39) – the laws of physics are the same for all observers in all frames of reference

Ionisation energy (p.90) – the energy level at which an electron escapes from the electric field of the nucleus

Irradiance (p.87) – the power per unit area of a radiation

Isotope (p.67) – isotopes of an element have the same number of protons but different numbers of neutrons; these isotopes have identical chemical properties but different nuclear reactions

LED (p.133) – a light-emitting diode; a semiconductor device that gives out photons when current passes through it

Leptons (p.63) – light particles that are themselves fundamental

Line spectra (p.91) – lines with certain frequencies emitted by an element

Lost volts (p.111) – the potential difference across the internal resistance

Magnetic field (p.55) – the area of space around a magnet

Magnitude (p.12) – the size of something

Maxima (p.77) – points of constructive interference

Meson (p.64) – a particle made up of two quarks

Minima (p.77) – points of destructive interference

Momentum (p.20) – the mass of an object multiplied by its velocity

Muon (p.40) – a subatomic particle with a short half-life

Negligible (p.36) – so small that it can usually be ignored

Neutrino (p.63) – one of the leptons, first discovered by studying beta decay

Newton's Third Law of Motion (p.25) – for every action there is an equal but opposite reaction

Normal (p.95) – a line drawn at right angles to a surface

n-type semiconductor (p.132) – semiconductor material in which the majority of charge carriers are negative

Ohm's Law (p.106) – the potential difference across a component is proportional to the current flowing through it, usually given by the equation $V = IR$

Open circuit (p.113) – a circuit in which no current is drawn

Optical density (p.95) – a measure of the way light interacts with a material as it passes through it

Particle accelerator (p.57) – a large machine used to accelerate particles to high speeds before colliding them together

Path difference (p.77) – the difference in distance from one source to a point and the other source to the point

Peak (p.106) – the maximum current or voltage

Peak wavelength (p.50) – the wavelength in a star's spectrum that has the highest irradiance

Percentage uncertainty (p.138) – the absolute uncertainty divided by the measurement multiplied by 100%

Period (p.108) – the time taken for one complete cycle or oscillation

Perpendicular (p.13) – at right angles

Phase (p.75) – how we describe the relative position of a wave

Photoelectric effect (p.84) – in the photoelectric effect, electrons are emitted from solids, liquids or gases when they absorb energy from light

Photoelectron (p.93) – an electron ejected from a material by the photoelectric effect

Photon (p.85) – a quantum of electromagnetic radiation

Planck's constant (p.85) – the constant of proportionality between the energy of a photon and its frequency, with a value of 6.63×10^{-34} J s

Point source (p.87) – where the light from a point spreads out evenly in all directions

Positron (p.63) – the antiparticle of the electron

Potential difference (p.54) – the number of joules per coulomb; often called voltage between two points

Projectile (p.30) – an object that has both a horizontal and vertical component to its motion

p-type semiconductor (p.132) – semiconductor material in which the majority of charge carriers are positive

Quarks (p.63) – the fundamental particles that make up the hadrons

Random uncertainty (p.138) – the uncertainty associated with repeated readings of the same quantity

Rest (p.2) – not moving; stationary

Recessional velocity (p.46) – the velocity at which a distant object is moving away from the Earth

Red shift (p.45) – the spectra of all distant objects are shifted towards the red end of the spectrum; the faster an object is travelling away from the observer, the more its spectrum is redshifted

Refract (p.95)– when a wave changes speed, wavelength and direction as it moves from one material to another

Refractive index (p.95) – the ratio of the speed in air to the speed in the material

Root mean square (r.m.s.) (p.107) – the r.m.s. AC voltage is equivalent to the DC voltage

Satellite (p.33) – an object in orbit around another

Scalar (p.7) – a quantity that has magnitude (size) and a unit

Semiconductor (p.130) – a material that will conduct under certain conditions; in semiconductors the gap between the valence band and the conduction band is much narrower than in insulators

Short circuit (p.113) – a path with zero electrical resistance

Solar cell (p.133) – a semiconductor device that produces a voltage when photons enter the semiconductor junction

Spontaneous (p.68) – happens naturally; in nuclear fission an isotope that decays without being bombarded by neutrons

Systematic uncertainty (p.139) – an uncertainty caused by the system, such as a meter not being calibrated correctly

Tau (p.63)–one of the leptons

Terminal potential difference (p.111) – the voltage measured across the terminals of a power supply

Terminal velocity (p.11) – the velocity of an object when the driving force acting on it is balanced by the friction force

Threshold frequency (p.85) – the minimum frequency of a photon which causes electron emission

Total internal reflection (p.99) – when all light is reflected internally within a transparent material

Transparent (p.101) – allows light to pass

Unbalanced forces (p.1) – forces that cause an object to accelerate

Valence band (p.130) – the highest occupied band in insulators; in insulators this band is full

Vector (p.3) – a quantity that has magnitude (size), a unit and a direction

Velocity (p.1) – the rate of change of displacement

Voltage (p.58) – another name for potential difference

Wavelength (p.98) – the minimum distance for a wave to repeat itself

Work done (p.17) – a measure of the energy transferred or converted during a process

Work function (p.86) – the minimum energy required to eject an electron from its parent atom

Zero potential energy (p.84) – an electron has zero potential energy in a nucleus' field when it has escaped from the field; this happens when ionisation takes place